GEOLOGICAL SOCIETY SPECIAL REPORT NO. 18

SERIES EDITOR: P. F. RAWSON

D1741541

Geophysical Logs in British Stratigraphy

Alfred Whittaker
Douglas Watson Holliday &
Ian Edwin Penn

1985
Published for
The Geological Society by
Blackwell Scientific Publications

Published for the Geological Society
by Blackwell Scientific Publications
Osney Mead, Oxford OX2 0EL
8 John Street, London WC1N 2ES
23 Ainslie Place, Edinburgh EH3 6AJ
52 Beacon Street, Boston, Massachusetts 02108, USA
667 Lytton Avenue, Palo Alto, California 94301, USA
107 Barry Street, Carlton, Victoria 3053, Australia

British Library Cataloguing in Publication Data
Whittaker, A.
Geophysical logs in British stratigraphy.—
(Special report, ISSN 0309-670X; no. 18)
1. Rocks, Sedimentary   2. Geology, Stratigraphic
3. Geophysical well logging
I. Title   II. Holliday, Douglas Watson   III. Penn,
Ian Edwin   IV. Geological Society of London
V. series
552′.5 QE471

ISBN 0-632-01488-1

Distributors

USA and Canada
Blackwell Scientific Publications Inc.
PO Box 50009, Palo Alto
California 94303

Australia
Blackwell Scientific Publications
(Australia) Pty Ltd
107 Barry Street
Carlton, Victoria 3053

Printed in Great Britain
at the Alden Press, Oxford

Preface

This report provides a demonstration that for Britain, as for most of the rest of the world, the procedures of stratigraphic correlation have changed out of recognition in the last 30 years. No longer is stratigraphy dependent on the contingency of surface exposures or mines, but the world's sedimentary basins have been probed by many tens of thousands of boreholes and electric logging in its various forms developed to produce a continuous physical record of formations from surface downwards in considerable detail.

On continental shelves and on land the great majority of geologists involved in stratigraphic correlation now rely almost totally on electric logs—and regrettably some at least rarely handle a rock.

In England the first electric logs (self-potential and resistivity) were run in BP's Portsdown and Henfield boreholes in 1937. Regular electric logging of oilfield tests and producing wells followed and became standard after the war, for example in the East Midlands oilfields. The purpose was, however, primarily for identification of potential reservoir sandstones and their fluid contents: coring for biostratigraphic age data was not a high priority and was minimal.

The work of the present authors marks a major step forward. The Deep Geology Research Group of the BGS undertook the investigation of the major onshore sedimentary basins of Britain, accumulating all available well data (much of which was confidential) and supplementing this with information from continuous coring operations to assist correlation and calibration of electric logs with full biostratigraphic control. The result, herewith, is a more complete account of the use of geophysical logs in the stratigraphy of British sedimentary rocks than has ever previously been available.

It was the widespread application of electric logging in North America which emphasized the need for distinguishing lithostratigraphy from biostratigraphy. These two methods are fundamentally different and contingencies of diachronism, facies change and unconformities may leave their results difficult to match. But the UK stratigraphic record, long known locally in very great detail, shows a surprisingly high degree of correlation between the physical and biological. It remains no less necessary to distinguish whether subdivisions are litho- or bio-stratigraphically based but the degree of correlation probably supports L. J. Wills' postulate that the British stratigraphic column consists to a significant degree of lacunae.

This report, published in the BGS's birthday year, will add greatly to the application of well logs in the interpretation of British stratigraphy and to understanding the development of our sedimentary basins through time. Those concerned are to be warmly congratulated.

February 1985

SIR PETER KENT, FRS
Past Chairman, NERC

Geol. Soc. London, Special Report, No. 18, 1985, 74 pp.

Contents

SUMMARY

The commonly available geophysical downhole logs are reviewed and some of their geological uses described. Lithological determination from logs can be achieved simply, by eye, or by using a computer; examples from British cored and logged borehole sections are presented. In combination with cuttings information, log data can provide much information of value in the fields of, for example, sedimentology, structural geology, shale compaction and depth of burial. Geophysical logs are particularly useful in stratigraphical correlation. Log data are not common, as yet, from the Lower Palaeozoic and older systems, but an increasing body of data from the British Upper Palaeozoic formations shows considerable stratigraphical consistency from place to place. The Permian and Mesozoic formations are extensively drilled and logged; a remarkable degree of correlation is possible within, and between, sedimentary basins. Examples of released or freely available log data are used to characterize the major stratigraphical units of the onshore British area.

1. INTRODUCTION

Although many earth scientists regard geophysical downhole logging as a very recent development, in fact the technique has quite a long and honourable history. The early development of the method owes much to the dedication and insight of the Schlumberger brothers, Conrad and Marcel, whose name is practically synonymous with the business of taking physical measurements downhole.

Between 1912 and 1926 Conrad Schlumberger, at first alone and then with a small number of colleagues, invented a surface-operated method of mineral prospecting by gathering electrical measurements which gave information on the geometrical and physical structure of subsurface formations (Allaud & Martin 1977). The impetus for early research into the applicability of electrical prospecting techniques came from the search for metalliferous mineral deposits; at the time, electrical conductivity appeared to be a much more significant geophysical parameter than density or magnetic properties.

As early as 1921 the Schlumberger team had attempted to take electrical resistivity readings in an exploratory borehole in the Bessèges coal basin. The records, taken over a metre or so at the bottom of the borehole, apparently reflected variations in the nature of the geological formations. However, it was 1927 before Conrad Schlumberger produced a note entitled 'Electrical research in boreholes' outlining the principle of the new method. At first the technique was known as 'electrical coring' (carottage électrique) because of the analogy with 'mechanical coring' which referred to the analysis of rock samples taken while drilling. Since 1933 the process has become known as 'electrical logging'.

The first electrical logging operation was undertaken in 1927 at Pechelbronn, Alsace, with primitive equipment and only 550 m of wire, in a borehole around 460 m deep. Measurements were not carried out continuously but at 1 m stations downhole. With practice the loggers could measure about 50 stations per hour. After the measurements were plotted it became clear that the variations in electrical resistivity corresponded to important lithological changes; eventually it became possible to correlate peaks and troughs of electrical resistivity in boreholes over a wide area. Comparison of the log data with cores in sufficient boreholes showed that accurate lithological identification could be achieved and logging gradually began to replace coring.

The subsequent history of downhole logging and the associated interpretation philosophy divides into several phases. By the early 1930s the spontaneous potential log had been tried successfully, and refined resistivity techniques, together with a device to measure the dip of strata, were developed. Further advances in the late 1930s included short and long normal logs, tools to measure downhole temperature, well drift and formation dips, plus the development of techniques of sidewall sampling and casing perforation. Thus, in addition to making subsurface correlations and distinguishing between shales, compact rocks and permeable or porous formations, the logging industry developed the important facility of sampling the subsurface after actual drilling operations had ceased.

After 1945 new resistivity logs were introduced including the micrologs and the deep resistivity logs (induction log, laterolog). To some degree this new spurt of activity was determined by the post-war thrust forward in hydrocarbon exploration. Early trials in the late 1930s of a gamma ray log had developed by 1946 into a fully operational system which was of immense value to the oil industry because logging could be carried out through steel casing and hidden formation boundaries delineated. In addition, this period saw the development of porosity logging with the introduction of the neutron log, the gamma gamma or density log and the continuous velocity log. Parallel development of the caliper log and improvements in other services permitted the addressing of many oilfield and related geological/petrophysical problems in the middle to late 1950s.

New downhole tools have come into operation since the late 1950s but other developments also have had an important impact on geophysical logging. One such is the combination of tools into one sonde to reduce logging time. Another is automatic computer data processing to give instant output and even interpretation in the field. For the future one may expect the development of hostile environment logs for use in very deep boreholes where formation temperatures and pressures are great, enhancement of resolution in conventional logging, greater use of the computer, the resolution of microstructure and improved lithological determination.

Most of these state-of-the-art developments are primarily of interest to the oil geologist and for the direct, downhole location and evaluation of mineral deposits. The purely geological or stratigraphical uses of downhole geophysical logs have been rather neglected, although they are of vital importance in the fields of stratigraphical correlation, lithological determination, recognition of sedimentary cycles, subsurface sedimentology and compaction and burial history. Enough data are now available from research-generated boreholes

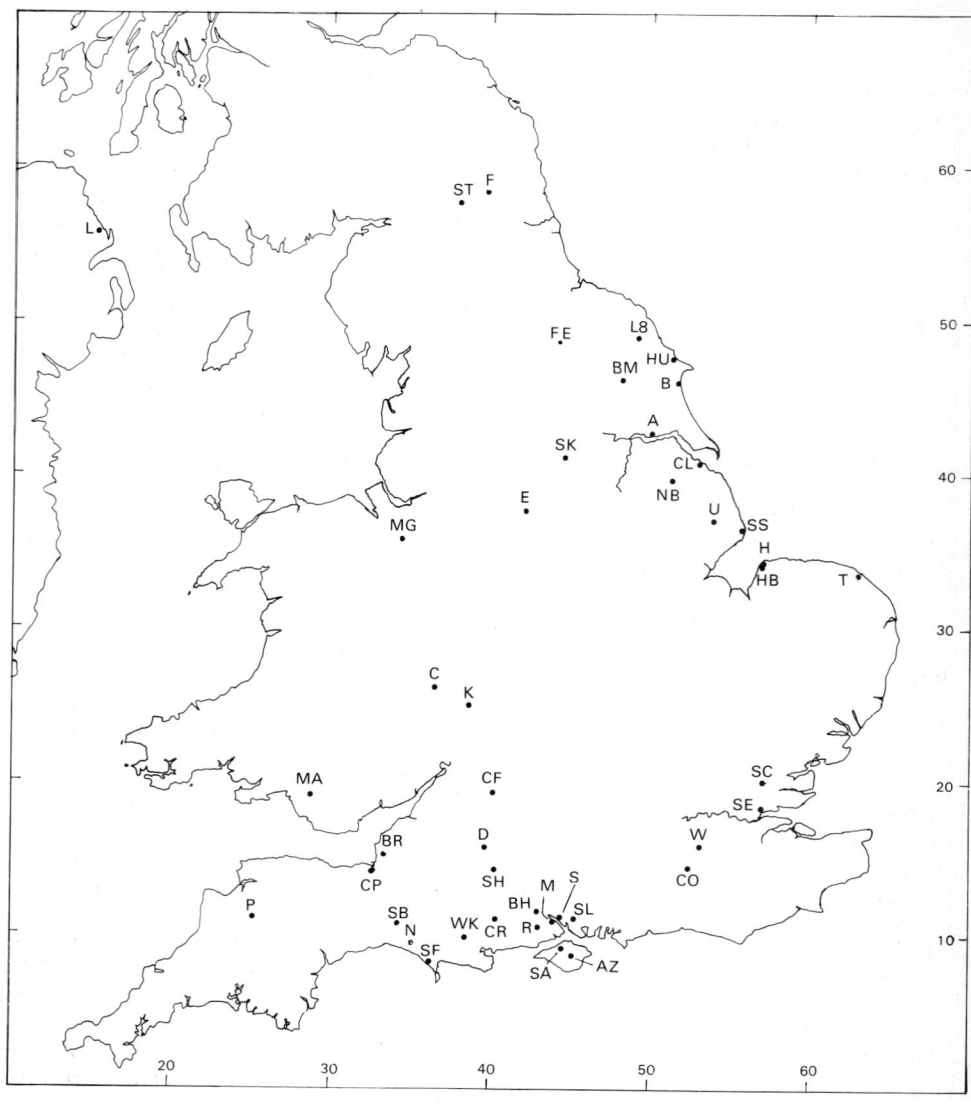

FIG. 1. Location of boreholes illustrated.

A	Alandale	[TA 0007 2584]	L8	Lockton 8	[SE 9106 8959]
AZ	Arreton 2	[SZ 5320 8580]	M	Marchwood	[SU 3991 1118]
B	Barmston	[TA 1545 6062]	MA	Maesteg	[SS 8528 9245]
BH	Bunker's Hill	[SU 3038 1498]	MG	Milton Green	[SJ 4374 5692]
BM	Brown Moor	[SE 8126 6203]	N	Nettlecombe	[SY 5052 9544]
BR	Burton Row	[ST 3356 5208]	NB	Nettleton Bottom	[TF 1252 9820]
C	Collington	[SO 6460 6100]	P	Petrockstow 1B	[SS 5202 1041]
CF	Cooles Farm	[SU 0164 9213]	R	Ramnor Inclosure	[SU 3114 0475]
CL	Cleethorpes	[TA 3024 0709]	S	Southampton	[SU 4156 1202]
CO	Collendean Farm	[TQ 2480 4429]	SA	Sandhills	[SZ 4570 9085]
CP	Cannington Park	[ST 2479 4011]	SB	Seaborough	[ST 4348 0620]
CR	Cranborne	[SU 0341 0907]	SC	Stock	[TL 7054 0045]
D	Devizes	[ST 9603 5699]	SE	Stanford-le-Hope	[TQ 6965 8241]
E	Eyam	[SK 2096 7603]	SF	Seabarn Farm	[SY 6263 8054]
F	Ferneyrigg	[NY 9579 8364]	SH	Shrewton	[SU 0314 4199]
FE	Felixkirk	[SE 4839 8569]	SK	South Kirkby	[SE 4546 1092]
H	Hunstanton 1	[TF 6923 4270]	SL	Shamblehurst Lane	[SU 4927 1456]
HB	Hunstanton BGS	[TF 6857 4078]	SS	Skegness	[TF 5711 6398]
HU	Hunmanby	[TA 1301 7588]	ST	Stonehaugh	[NY 7899 7619]
K	Kempsey	[SO 8609 4933]	T	Trunch	[TG 2933 3455]
L	Larne 2	[D 4070 0226]	U	Ulceby Cross	[TF 4140 7385]
		(IRISH GRID)	W	Warlingham	[TQ 3476 5719]
			WK	Winterborne Kingston	[SY 8470 9790]

cal determination are the gamma ray, sonic, formation density and neutron. Additionally, in many parts of the world, SP logs are extensively used for this purpose. However, in British boreholes there is commonly insufficient salinity difference between the mud filtrate and formation waters to produce characterful logs which can both define lithology and positive sharp bed boundaries. Our experience is that the gamma ray log is much more valuable in this respect than the SP log and is generally to be preferred. However, in older boreholes, no gamma ray log may be available and an SP must suffice. Similarly, sonic, density and neutron logs are not always available and recourse to resistivity logs for lithological determination is then necessary.

The gamma ray log is perhaps the single most important log used in stratigraphical correlation. Nearly all of the gamma radiation encountered in the earth is emitted by radioactive isotopes of potassium, uranium and thorium. In sedimentary formations the log can be used as an indication of shale content because the radioactive elements, notably potassium, tend to concentrate in clays and shales. Thus argillaceous rocks generally have high radioactivity values, whereas limestones and sandstones commonly give low readings. There are many exceptions to this, however, sandstones with feldspars and/or mica and/or radioactive heavy minerals being notable examples. Another use of the gamma ray log is the distinction of potash minerals in evaporite sequences (approximately 15 API units for 1% of K_2O). More recent developments enabling differentiation of gamma rays of different energies (spectral gamma log) allow the separate identification of potassium, thorium and uranium components and thus permit detailed determination of clay mineralogy.

The sonic log values, measuring the speed of compressional sound waves, respond mainly to both lithology and intergranular porosity. Values (in microseconds per foot) for common rock types are sandstone 51–56, limestone 47.5, dolomite 43.5, anhydrite 50 and halite 67. Shales and particularly coals give variable responses generally in excess of the above. Values from the first three rock types are increased by the presence of significant pore spaces. Nevertheless the sonic log, particularly in tandem with the gamma ray log, has proved to be a sensitive lithological indicator.

Although the prime use of the density log is the indication and definition of porosity and determination of the nature of the pore fluids, it is also a powerful lithological indicator because of the wide range of rock densities (g/cm^3) naturally encountered, e.g. coal 1.2 to 1.8, halite 2.05, sandstone up to 2.65, limestone up to 2.71, dolomite up to 2.87, anhydrite 2.98. The density log is of particular use in carbonate/evaporite sequences and in the detection of coal seams.

The neutron log primarily is another porosity tool. Nevertheless its ability to distinguish shales (high porosity) from limestone or sandstone (moderate to low porosity) makes it a valuable indicator of lithology, especially where sonic and density logs are not available. It becomes especially valuable in the definition of complex multi-mineral rocks as discussed more fully below.

The SP log detects beds invaded by the drill mud and, in combination with the caliper log, can be used to distinguish relatively more permeable beds such as sandstone and limestone from shales. Against the shale part of a sequence, the log tends to make a straight line known as the 'shale base line'; against a permeable sandstone the log moves from the 'shale base line' to a 'sand line' where thick enough beds are present to achieve a constant deflection.

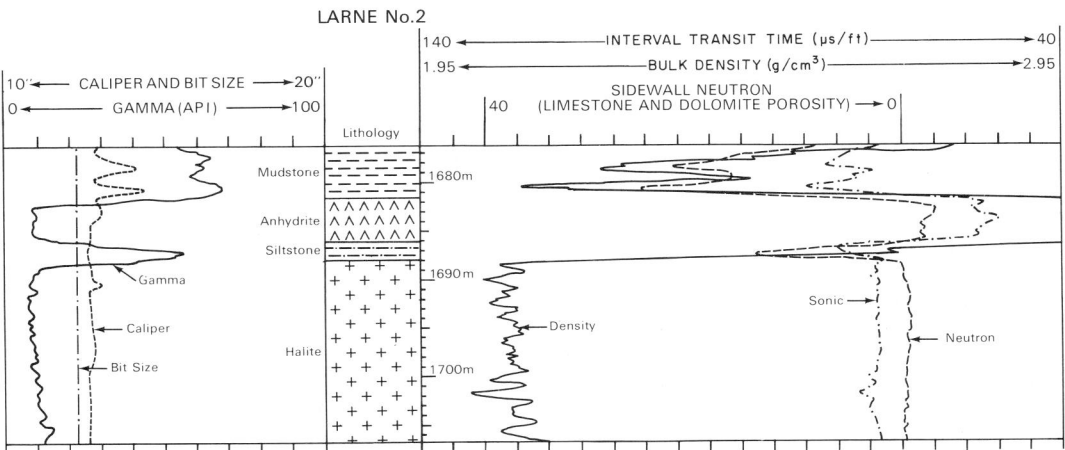

FIG. 2. Log signature and lithology of a Permian evaporite sequence.

Resistivity logs are useful in picking out tight, low porosity beds, e.g. some sandstones and carbonates, and evaporites, which because of their low formation water content resist the passage of electric currents. More porous sandstone and carbonates can be distinguished from shales with more difficulty. It should be remembered that many resistivity logs are designed to investigate the zone invaded by drilling mud and are not intended for primary lithological identification.

If formations were comprised of monomineralic layers then it would be a relatively simple matter to compare values obtained from geophysical logs with those determined experimentally and so deduce lithology. For example, the Larne No. 2 Borehole (Penn *et al.* 1983) encountered 4 m of anhydrite almost immediately overlying (*c.* 1688 m depth) a very uniform 113 m of halite, both of Permian age. Examination of the sonic log (Fig. 2)

shows the interval velocities of the anhydrite (50–58 μs/ft) and the halite (68 μs/ft) in a true gauge hole to be almost exactly those determined experimentally (Fig. 3). However, similar interval velocities may be obtained, for example, from clean limestone with 1.5–6% and 15% water-filled porosity, respectively. If the lithology is unknown in the first place, then the elimination of such a possibility may be made by examination of the density log. This shows the upper bed to yield a formation density greater than 2.95 g/cm^3 and the lower bed to yield a formation density of 2.10 g/cm^3, which in combination with the interval transit times indicate very pure anhydrite and halite, respectively (Fig. 3). Examination of the gamma ray log shows both evaporites to yield values of less than 10 API units and confirms the purity of the evaporites.

A comparable case of similar simplicity exists in Coal Measures sequences. Core cut in Westphalian

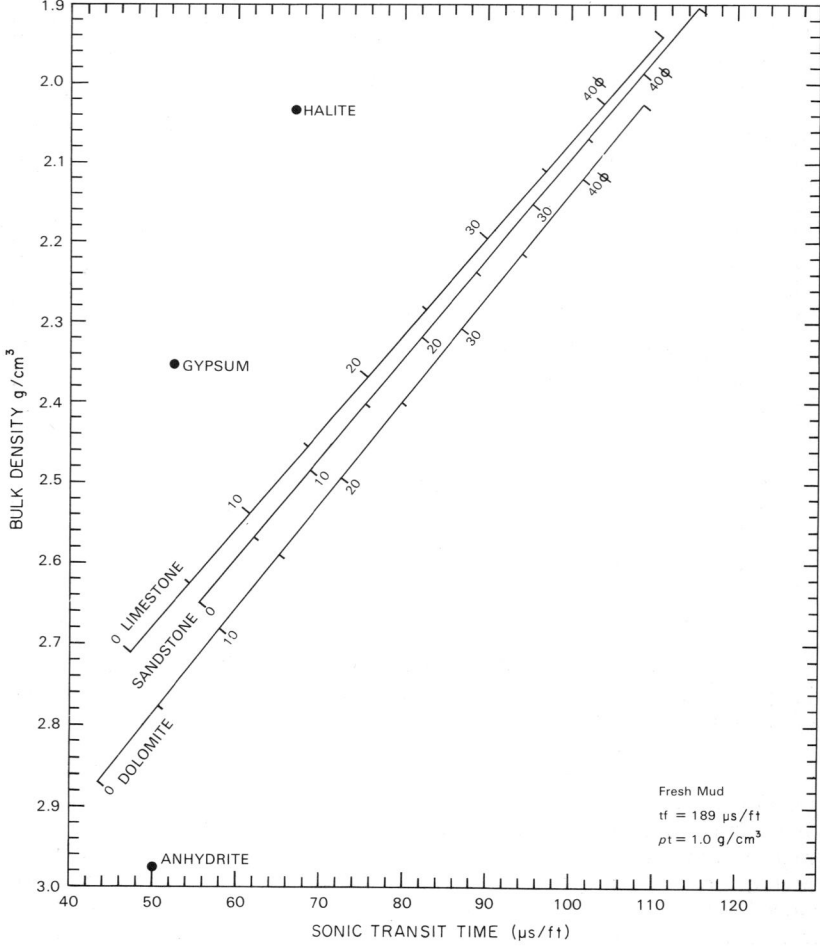

Fig. 3. Density–Sonic cross-plot (Schlumberger 1979, chart CP–7).

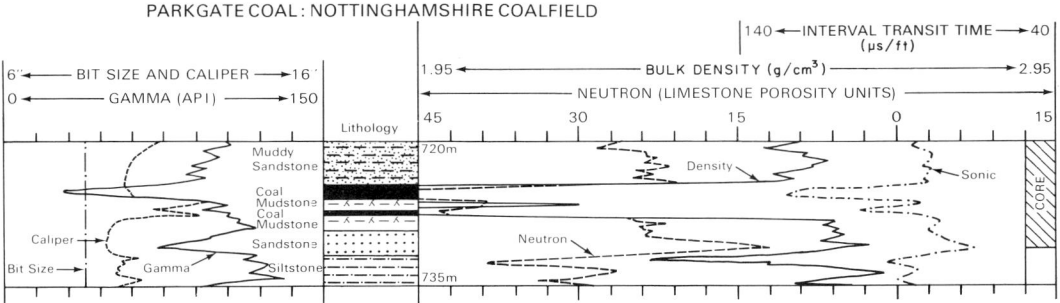

FIG. 4. Log signature and lithology of a Coal Measures sequence.

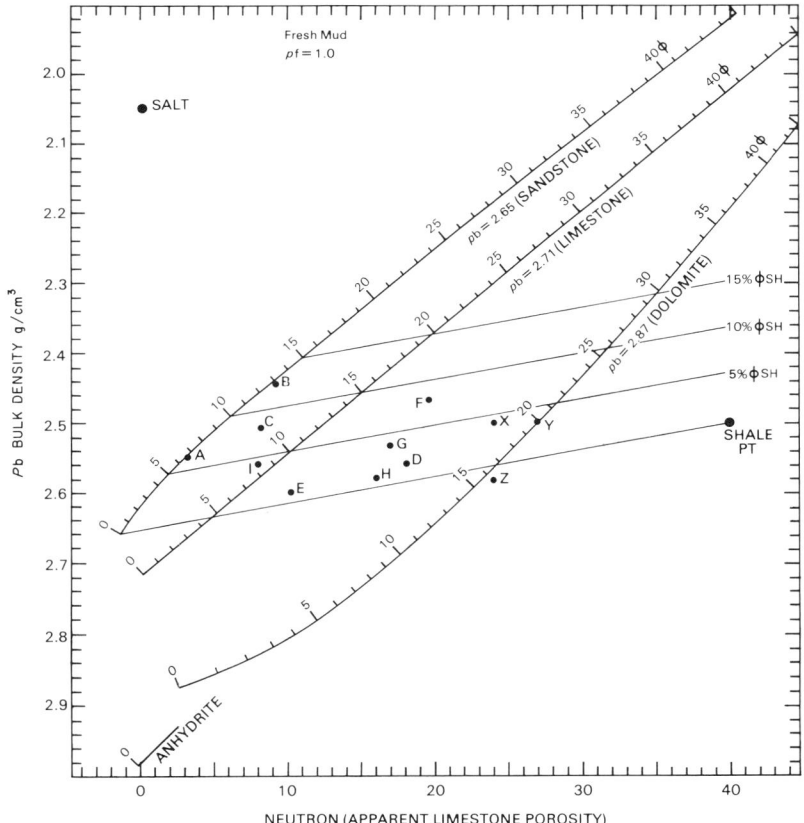

FIG. 5. Density–Neutron cross-plot (Schlumberger 1979, chart CP–1c). See Fig. 8 for location of data-points A to I and X, Y, Z.

'A' strata proved the existence of a major and minor coal seam each underlain by seat-earths and occurring within a succession of siltstones and more or less muddy sandstones (Fig. 4). The coals are easily identified by their low gamma ray values (25 API) and high interval transit times of 135 μs/ft combined with the exceptionally low bulk density (1.33 g/cm³) and high neutron porosity values (53 limestone porosity units).

These simple examples demonstrate the resolution gained by using geophysical log data in combination rather than considering each singly. By plotting interval transit time against bulk density or the latter against neutron 'porosity' it is possible to locate the position of different minerals uniquely. By varying the porosity of each and assuming a standard fluid filling and saturation, namely 100% water-filled porosity, it is possible to translate the mineral points into unique curves which describe both mineralogy and porosity. Lithology expressed as mineral combinations may be estimated by interpolation between the curves, and water-filled porosity by interpolation along their length (Fig. 5).

Such two-dimensional cross-plots, however, do not produce unique solutions by themselves. For example, a pure limestone (calcite) may yield the same neutron and density values as a sandy dolomite (Fig. 5). The latter would be distinguished from the former by considering the appropriate sonic/density relationship (Fig. 3). The more complex the lithology (and the more complex any contained fluids) the more log data are needed to

evaluate the formation. In addition, such two-dimensional 'cross-plots' summarize data over individual stratigraphical intervals but do not conveniently display lithological (and porosity) variation in a complete geological succession. A convenient way to represent the cross-plot data or combined geophysical log data in a stratigraphical sequence is to overlay standardized log traces. It is particularly useful to overlay the bulk density and neutron porosity log traces, standardizing them so that the zero porosity value corresponds with the bulk density (2.71 g/cm^3 usually taken for simplicity as 2.70 g/cm^3) of pure calcite and to scale the density values appropriately. This results in two log traces of a pure limestone sequence coinciding at a limestone porosity value corresponding to the water-filled porosity of the formation.

For example, the 'Great Oolite' encountered in the Marchwood No. 1 Borehole (Fig. 6) was

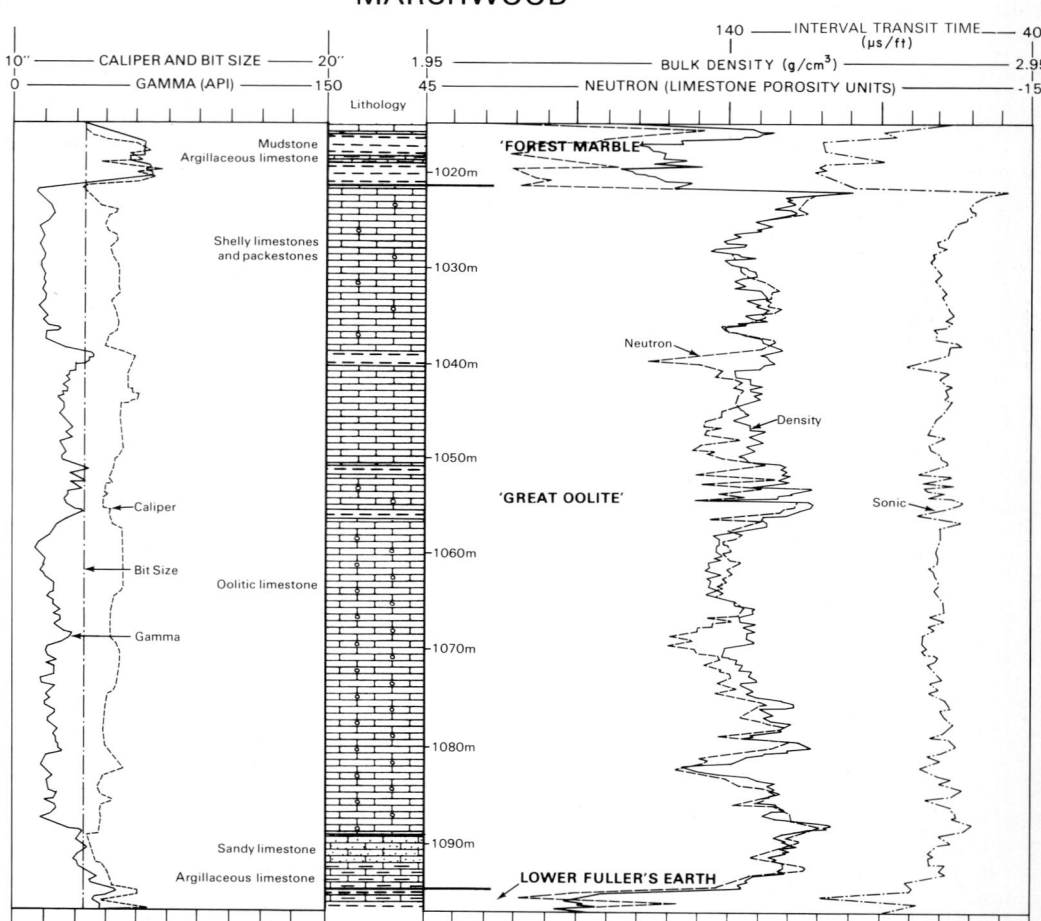

FIG. 6. Log signature and lithology of Upper Great Oolite Group at Marchwood.

water-filled and, as confirmed by the low API values of the gamma ray trace, found to be relatively clean (Whittaker 1980c). The neutron and density log traces are approximately coincident and indicate water-filled porosity in the range 9–18%. Mudstone predominates above and below the 'Great Oolite'; the gamma ray values increase accordingly and the density and neutron log traces are at values and separation consistent with this lithology found in the cuttings. The Southampton No. 1 Borehole, located 1.8 km to the north-east (Thomas & Holliday 1982), proved a closely comparable sequence to that at Marchwood, save that the strata between 1030 m and 1040 m (Fig. 6) at the latter may be absent at Southampton. The log traces are similar in both sections but the gamma ray values are somewhat higher in the limestones at Southampton (Fig. 7). Both density log traces have similar profiles but the most striking difference of the overlaid log traces is that the neutron porosity log trace at Southampton, while also maintaining a similar general profile, is displayed exceptionally consistently some nine limestone porosity units higher than the density log trace. Plotting of these data on a density/neutron cross-plot suggests that the Southampton sequence is strongly dolomitic by comparison but lies within a comparable range of water-filled porosity. Porosity derived from the sonic log gives similar results in both cases and indicates the probable absence of secondary porosity.

A more complex lithological sequence is known from cores taken in the Sherwood Sandstone and contiguous strata of the Southampton No. 1 Borehole. Porosity measurements were made on some of the sandier cores so that an independent check on both lithology and porosity is available (Thomas & Holliday 1982). The major lithostratigraphic units are readily picked out by the geophysical log signatures (Fig. 8). The separation in the density/neutron log traces at 1735.7 m (A) and 1730 m (C) is consistent with relatively clean and pure sandstones of 6% and 10.5% porosity respect-

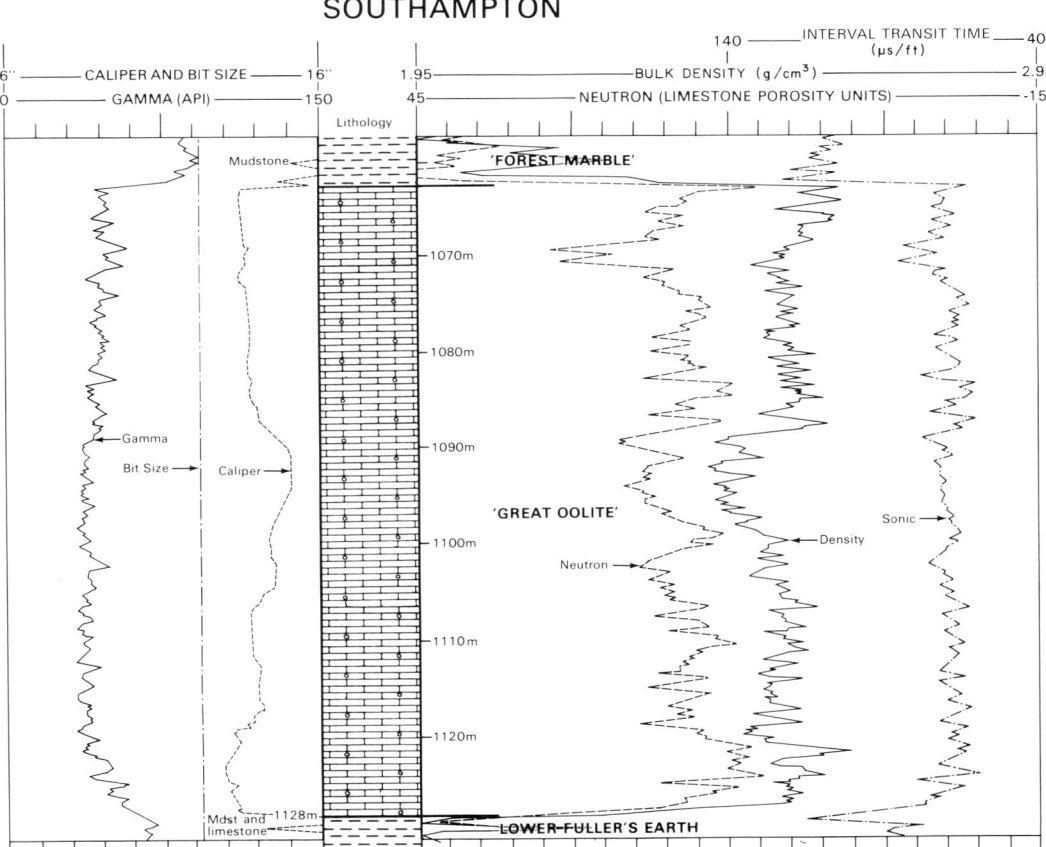

FIG. 7. Log signature and lithology of Upper Great Oolite Group at Southampton.

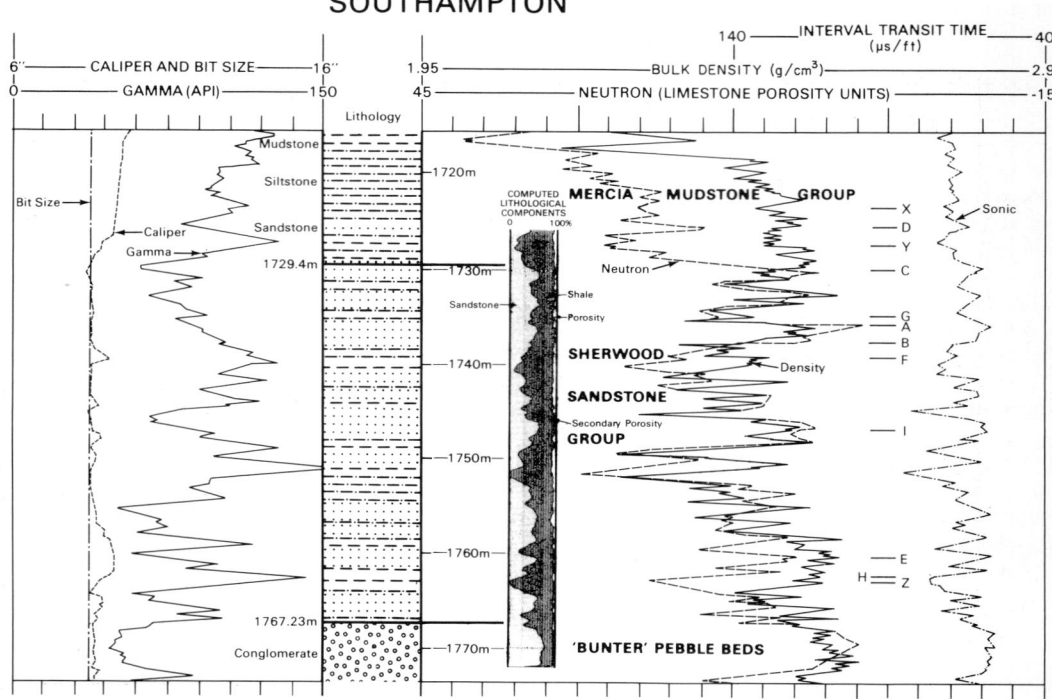

FIG. 8. Log signature and lithology of Sherwood Sandstone and contiguous strata at Southampton, including computer constructed lithological section.

ively (Fig. 5). The lower part of the sandstone, with base at 1737.7 m (B), is more porous (12.5%). Laboratory determinations of porosity for these sandstones are 10.5% for that at 1730 m and 3.03–8.65% for that at 1737 m. The sandstones at 1726 m (D), 1747 m (I) and 1760 m (E) do not give values of bulk density and neutron porosity consistent with pure sandstones. It is not clear from the geophysical logs whether they are dolomitic sandstones with porosity values of 13.5%, 9% and 9% respectively or whether they comprise an admixture of sand and mud.

Similar ambiguity exists for the siltstones at 1739 m (F) and 1735 m (G) and the lower part of the sandstone at 1762.5 m (H). It will be seen that these horizons lie between the sandstone line and the positions of the silty mudstone horizons at 1723.5 m (X), 1727.5 m (Y) and 1763 m (Z) on the density/neutron cross-plot (Fig. 5).

There is a limit to the number of logs that can be conveniently handled by visual examination and cross-plotting techniques. It is possible, however, to consider three logs simultaneously by cross-plotting ratios of each of two against the values of the third and identify characteristic mineralogy/litho-

logy in the usual way. Fortunately the burden of basically simple but laborious calculation has nowadays been relieved almost entirely by the advent of electronic computing, particularly where interactive facilities are available which allow the geologist to control the iterative solution of equations. In addition, the practice of digitizing the log data and storing it on magnetic tape allows many more levels in a stratigraphical sequence to be analysed, and standard graphics facilities allow display in the form of a conventional strata log. Indeed the portability of modern computing equipment allows formation evaluation to be carried out at the drill-site if required.

Care, however, has to be taken in handling such digitized data. For example, in dealing with a part of a sequence, the appropriate header record, which contains much important drilling information, may become detached from the log data and ignored. Perhaps the most important caveat is that the computer always finds a solution to the equations and, in the last analysis, it depends on the geologist to ensure that these are geologically sensible.

An automated analysis of the stratigraphical

SOUTHAMPTON

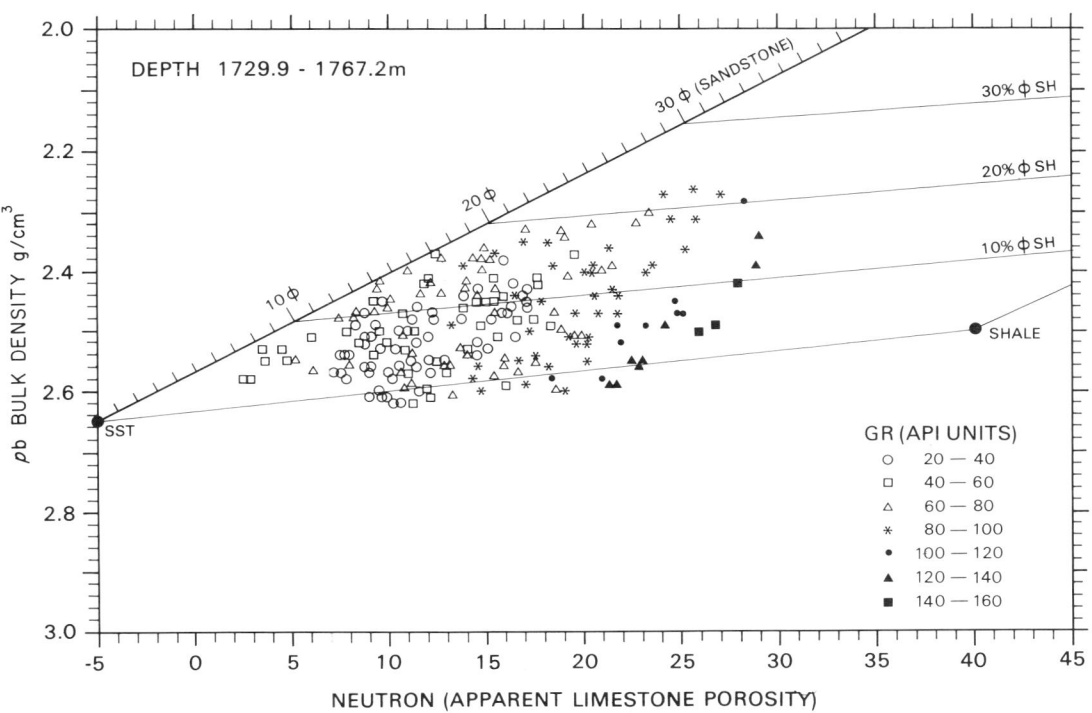

FIG. 9. Density–Neutron cross-plot (including shaliness indicator). Data of Fig. 8.

interval of Fig. 8 enables several hundred levels to be sampled and conveniently plotted on the familiar density/neutron cross-plot. In addition, the values may be scaled according to the degree of shaliness of each as indicated by, for example, their gamma ray values (Fig. 9). It is then possible to display the entire section almost immediately in terms of its lithological components which in the simple analysis shown comprise sandstone and mudstone (Fig. 8). As can be seen, corresponding porosity and possible secondary porosity values may also be displayed. More sophisticated analysis, not attempted here, would assess the extent of the calcareous and dolomitic nature of the sequence.

Use of geophysical logs in sedimentology

Important sedimentological data can be determined from geophysical logs, particularly when these are used in conjunction with lithological and micropalaeontological analysis of cuttings and where the formation is known either from core obtained in nearby boreholes or from surface

outcrops. Such studies are widely employed in the petroleum industry as a tool in basin evaluation, as a means of deducing the nature and potential of proven reservoir formations, and in the siting of development wells. The methods used fall into two main categories. Much important sedimentological data can be gleaned from the routine examination of a standard log suite, in determining bed thickness, nature of lithological boundaries, and in the recognition of rhythmic lithological sequences (cyclothems). Secondly the use of a dip-meter may allow the scale, nature and orientation of bedding and cross-bedding to be deduced. By such methods detailed palaeoenvironmental reconstruction (integrated facies analysis) can be attempted.

Common lithological logs

Petrographical analysis of rock cuttings from an open-hole borehole reveals much to the sedimentologist about the lithologies penetrated, but rarely gives much detail regarding the nature of the interrelationships of the main strata types proved. Yet such knowledge is commonly essential if environmental reconstruction is to be attempted.

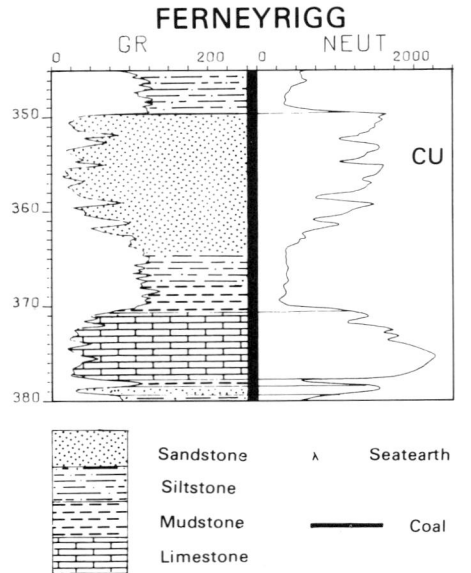
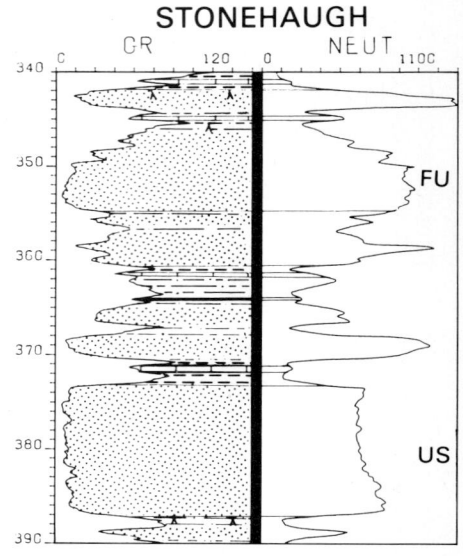

FIG. 10. Recognition of rhythmic sequences and nature of lithological boundaries. Ferneyrigg Borehole: typical 'Yoredale' cyclothem and coarsening-upwards clastic unit (CU), sandstone with gradational base. Stonehaugh Borehole: sandstones with both sharp and gradational boundaries, including sharply based fining-upwards unit (FU) and uniform sandstone with sharp top and base (US).

However, information of this kind is usually readily available from geophysical logs, though some caution in their interpretation is required.

Except in very poorly sorted or other exceptional rocks, the gamma ray log in clastic sequences may correspond closely to grain size profiles. For example, Fig. 10 shows how some simple, commonly occurring, sedimentary sequences may respond to the gamma ray and neutron tools. Both logs commonly allow the easy distinction of shales (mudstones and siltstones) from sandstones, or of shales from limestones. In such instances bed thickness is readily apparent, as is the nature of the bed contact which can be deduced from whether the log values change gradually (gradational boundary) or rapidly (sharp boundary). In many instances rhythmic lithological sequences can be recognized, e.g. fining and coarsening-upward cycles, Yoredale and evaporitic cyclothems. Further examples, similar to those shown in Fig. 10, are to be found in the figures illustrating the log response of typical British rock successions, as well as other commonly occurring sedimentological associations.

It should be stressed that other geophysical logs are useful in this context. For instance, density logs are important when working with evaporitic sequences. Indeed the gamma ray log, if used on its own, can lead to erroneous interpretations. For example, a sharply based channel sandstone might normally be expected to have a lower natural radioactive level than its underlying shales; however, if, as is quite common, the sandstone contains potash feldspars and/or radioactive heavy minerals and/or is heavily micaceous and/or contains numerous shale intraclasts near its base, then it will be less readily distinguished from the shale, and the sharp base may appear gradational from the gamma ray log. Similarly the sonic log may fail to indicate a sharp boundary where shale intraclasts occur.

Dipmeter

In recent years the dipmeter has become an essential tool in subsurface exploration, providing both structural and sedimentological data that would not otherwise be generally available from open-hole drilling. Much of the detailed specialized work stems from the identification of four basic, easily recognized, tadpole plot motifs (Fig. 11). By convention the various motifs are identified by log analysts using a standard colour code. A green pattern is a succession of dips of similar orientation and magnitude, and in the majority of cases is indicative of regional structural dip. Such patterns are common in argillaceous rocks or other thinly bedded or laminated sediments laid down in low-energy environments. Red (upward decreasing

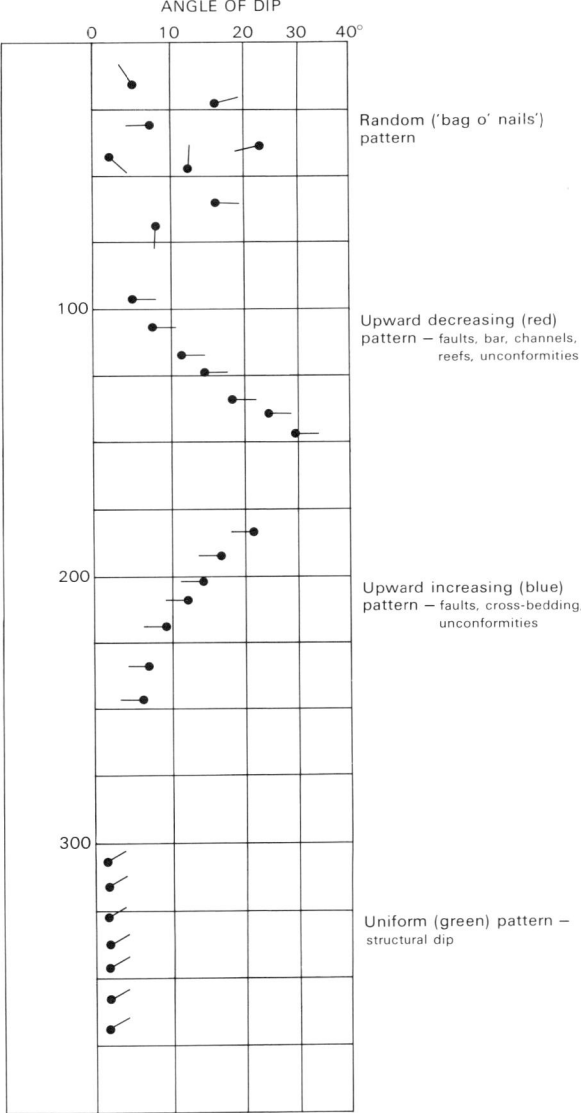

ANGLE OF DIP

Random ('bag o' nails') pattern

Upward decreasing (red) pattern — faults, bar, channels, reefs, unconformities

Upward increasing (blue) pattern — faults, cross-bedding, unconformities

Uniform (green) pattern — structural dip

FIG. 11. Simplified common dipmeter log motifs.

dips) and blue (upward increasing dips) patterns are commonly indicative of both faults and cross-bedding. Where a sedimentary origin can be inferred or demonstrated, information such as the size of the cosets and their nature and orientation can be used in environmental interpretation or for prediction of the subsurface extent of reservoir bodies. In sequences of mixed lithology it is possible to determine the structural dip from green patterns, and then by use of the computer to tilt back the red and blue patterns to their original orientations to give the direction of sediment transport. The recognition of faults has important sedimentological implications, not only because these may indicate missing, or repeated, strata and hence affect any sedimentological models proposed, but because it may be possible to infer from the dip-meter data that the faults were themselves active during sedimentation (growth faults). Clusters of randomly orientated tadpoles—'bag o'nails' pattern—can result from a number of causes. In many instances such patterns result from poor data or the inability of the computer to make proper correlation between the resistivity traces. A random pattern commonly indicates a structurally disturbed zone, due either to faulting or fracturing. However, in some cases the 'bag o'nails' pattern can result from sedimentological causes, most notably in thick massive bodies with little internal orientation, such as reefs, or where contemporaneous slumping has locally intensely deformed the bedding.

Integrated facies analysis

Even in areas of good surface exposure, detailed sedimentological analysis and unambiguous environmental reconstruction can prove difficult. Thus such studies are likely to prove even more problematical in open-hole boreholes. However, although the individual use of the main lithological and porosity logs, or the dip-meter, or analysis of cuttings, cannot lead to a full understanding of a sedimentary sequence, a more comprehensive picture may emerge from fully integrated studies of all three, with additional data from cores and surface exposures.

There is no single set pattern of approach to subsurface facies analysis as every case must be treated on its merits and limitations. However, a typical example might proceed in the following manner and pose these questions.

1. Cuttings examination
   a. What are the main lithologies present and in what proportion?
   b. What is the grain size, sorting, rounding, etc.?
   c. What important accessories of possible environmental significance are present, e.g. glauconite, carbonaceous matter, microfossils, shell fragments?
2. Log analysis
   a. Are the same rock types seen in the cuttings identifiable and in the same proportion?
   b. How thick are the main lithological units?
   c. Are the lithological boundaries sharp or gradational?
   d. Are fining-upwards and/or coarsening-upwards sequences recognizable? How thick are they?
   e. Are other rhythmic sequences present?
3. Dip-meter
   a. Does the tadpole plot show good correlation between the resistivity logs?
   b. What is the structural dip?
   c. Do any beds show cross-stratification? (blue or red patterns?) If so, how large and how numerous are the sets?
   d. Do the foresets have a preferred orientation?

Subject to this sort of analysis, fluvio-deltaic Westphalian (Coal Measures) rocks from the East Midlands area would give the following results.

1. Formation dominated by grey siltstones and mudstones, with sporadic fine-grained, slightly carbonaceous and micaceous, poor to moderately sorted sandstone. Medium-grained or coarser sandstone locally common but generally rare. Coal sporadically common. Shell debris indeterminate and rare.
2. Sediments form both fining-upwards and coarsening-upwards cycles, commonly capped by coal. Cycles only a few metres thick in main. Medium-grained sandstones in thicker (*c.* 10 m) fining-upwards sequences.
3. Argillaceous rocks give green patterns. Sandstones generally have blue patterns, cosets up to 1 m, randomly directed. Larger cosets in medium-grained sandstones.

Similar analysis of aeolian Lower Permian (Rotliegendes) sandstones from the southern North Sea might have the following features.

1. Fine- to coarse-grained sandstones, subangular to subrounded, frosted, well-sorted grains. Few other lithologies present and in small quantities. Accessories rare.
2. Thick uniform sandstone unit.
3. Large cosets (blue patterns), up to several metres thick. Foreset dip-direction variable but rose plot suggests overall movement from east.

The above examples are somewhat simplified and generalized, but are illustrative of the methods employed which in practice can be highly sophisticated. A number of studies have been published relating to rocks from the North Sea (e.g. Selley 1976, Parry *et al.* 1981).

Shale compaction and depth of burial

Examination of the sonic logs obtained from broadly uniform lithological sequences (e.g. Chalk, Coal Measures) reveals a general increase in velocity with depth. As a result the maximum depth of burial of argillaceous formations, the amount of uplift and the thickness of overlying strata removed by erosion may be estimated. This is possible because the compaction of argillaceous rocks is related to overburden or burial depth if the pore pressure is normal or hydrostatic. The degree of compaction can be determined from the sonic log because the sonic transit-time is a function of porosity in a uniform lithology (see Magara 1976). Such methods can be used only in rocks where the degree of compaction has been less than that necessary to reduce porosity to zero. Studies of this kind can provide data of great scientific and economic importance. Estimates of the amount and timing of maximum burial of argillaceous rocks, taken with known geothermal gradients and other palaeotemperature indicators, provide valuable clues regarding possible hydrocarbon generation and migration. Determination of the thickness of overlying beds now eroded away allows for estimates to be made of the former extent and thickness of younger formations and thus helps in palaeogeographical studies. Care should be taken to see that the sonic logs are borehole-compensated and that caliper logs are consulted to ensure that measurements of interval transit-times are not taken in parts of the borehole that were excessively over-gauge.

A number of differing methods have been proposed whereby burial history can be elucidated by use of the sonic log. One such, applied by Marie (1975), requires a plot, on normal graph paper, of average velocities in the Permo-Triassic Bunter Shale of the southern North Sea (= Permian Upper Marl of onshore areas) against the depth of the base of the formation, i.e. at the top of the Bröckelschiefer. The relative uniformity of the Bunter Shale/Permian Upper Marl, as judged from geophysical log analysis and from cuttings descriptions, and the basin-wide extent of the Bröckelschiefer marker, make these obvious choices for depth of burial studies. The velocities obtained are compared with a standard velocity curve (Marie 1975, Fig. 1) derived from boreholes in the southern North Sea judged to have had a 'normal compaction history', i.e. boreholes in areas of relatively complete sedimentation and minimal erosion in Mesozoic, Tertiary and Quaternary times. Comparison of the plotted value vertically to the main trend provides an estimate of the maximum depth of burial of the Bröckelschiefer/Bunter Shale interval, and thus also of the amount of later uplift and the thickness of overlying strata formerly present at the borehole site.

Figure 12 shows the results of applying such studies to the land area of eastern England, obtained from analysis of sonic logs derived from the Permian Upper Marls and the Coal Measures. A marked feature of the map is the inversion structure of the Cleveland Basin which shows both great Jurassic–Cretaceous subsidence and high post-Cretaceous (probably early Tertiary) uplift (Marie 1975, Kent 1980, Hemingway & Riddler 1982). These estimates of maximum post-Cretaceous uplift (*c.* 2.5 km) are in general agreement with the estimates of Hemingway & Riddler (1982, p. 185) based on mineralogical and petrological

criteria and with the work of Marie (1975) (1.8 km +) in offshore areas. As Kent (1980) has pointed out, these values are much greater than would be inferred (*c.* 1 km) from known and projected thicknesses. The results obtained support the evidence from surface geology for increasing post-Cretaceous uplift from the coast towards the Pennines. However, preliminary analysis of the data suggest that in the west of the region, where older rocks crop out, uplift is not as great as would have been expected from extrapolating known thicknesses of younger strata near the coast. Such results would be consistent with the generally anticipated thinning of Mesozoic formations from the coast to the Pennines.

4. LOG SIGNATURES IN BRITISH STRATIGRAPHY

The geological succession derived from log analysis and cuttings examination can be compared with established standards and a correlation made. Such correlations may be at varying levels of detail; for example, a borehole through the Carboniferous rocks of Nottinghamshire might prove four broad lithological groupings each 100 or more metres thick, viz.

Mudstones and siltstones, subordinate sandstone and coal	—COAL MEASURES
Sandstones, shale partings ⎫ Shales, minor sandstone ⎬	—MILLSTONE GRIT
Limestone, grey, massive	—CARBONIFEROUS LIMESTONE

More detailed analysis might allow the naming of individual beds and markers, e.g. Top Hard Coal, Vanderbeckei Marine Band, Chatsworth Grit, and provide a much more detailed correlation. Although these correlations are principally lithostratigraphical in nature, where thin persistent markers are recognized, such of necessity have chronostratigraphical significance, e.g. the base of the Vanderbeckei Marine Band is taken as the base of Westphalian B. Locally it has been established that minor lithological changes can be recognized over wide areas and may provide a finer subdivision that can be achieved by palaeontological study, e.g. Upper Jurassic clays of southern Britain (Gallois 1979). Since such lithological changes can be recognized by their geophysical log patterns, then the possibility arises, in these cases, of correlation to below ammonite zonal level in an open-hole borehole. Generally such precision is not possible, but broadly synchronous, basin-wide lithological changes commonly can be recognized and used as a

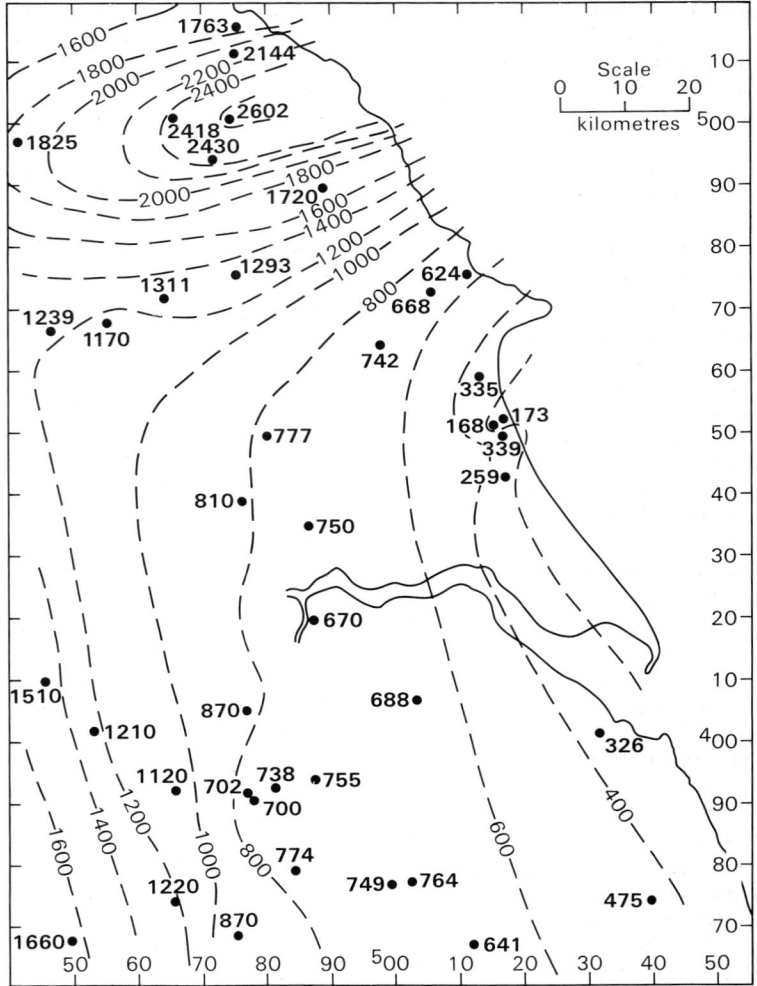

Fig. 12. Estimates of post-Cretaceous uplift in eastern England derived from study of sonic logs.
Based on unpublished work by Mrs J. M. Allsop and Dr G. A. Kirby.

basis for correlation. Examples from British stratigraphy are discussed and illustrated below.

Particular emphasis has been placed on cored sections since in such sequences the stratigraphical units are most precisely and accurately delimited. The long history of British onshore drilling accounts for the fact that many earlier boreholes were only primitively or poorly logged compared with the more recent offshore wells. This is counterbalanced, however, by the considerable precision of onshore British Phanerozoic stratigraphy. In the accompanying illustrations the gamma ray (GR) log is graduated in API units unless otherwise indicated, and the sonic log (interval transit time) in

microseconds per foot. BHCS indicates that the sonic log is borehole-compensated and SONL that it is not. Depths are in metres and the presence of core is indicated by black shading between the log traces.

Igneous, metamorphic and Precambrian rocks

The growing body of data from igneous, metamorphic and Precambrian rocks necessitates brief review here although in the strictly stratigraphical context, use is limited other than in metasedimentary sequences. Geophysical logs from selected Caledonian granites in northern Britain, acquired as part of the Hot Dry Rock geothermal energy programme, were described by Lee (1984). Six boreholes at Shap, Skiddaw, Cairngorm, Ballater, Mount Battock and Bennachie were logged to identify changes in rock condition, lithology and physical properties by comparison with short cored sections. Some of the logs were used to estimate the proportion of altered and jointed rocks which were characterized by low sonic velocity and high porosity; a wide range of physical properties was encountered in what previously were thought to be fairly uniform granites. Similar work was carried out by McCann *et al.* (1981) in boreholes drilled at Altnabreac, Caithness, in granite and Moine metasediments. They found that the most significant indicators of geological variations are the nuclear geophysical logs (gamma, gamma spectrometry, density, neutron) which respond more to geochemical changes in the rock fabric. Sonic and resistivity logs showed little variation in such a low porosity environment except where the rock mass was fractured or weathered. Of particular interest is the fact that logs from the Moine metasedimentary sequence showed more character than the others and exhibited a response similar to that found in unmetamorphosed sedimentary sequences; the diplog plot from the borehole compares well with information derived from the cores.

Farther south, rocks of Precambrian age have been logged only infrequently in boreholes. One example is the volcaniclastic ?Precambrian sequence proved at Kempsey (Whittaker 1980b), 5 km south of Worcester. Here, 680 m of tuffaceous and agglomeratic sandstones were drilled beneath the basal Permian unconformity which lies at a depth of about 2300 m. Although the sonic log shows fast velocities, in detail it is well-serrated and the gamma log is characterful, suggesting stratigraphical variation in the sequence; the dipmeter log shows well-defined dips between 10° and 40° (average 20°) to the SW.

Volcanic rocks have been geophysically logged, particularly in boreholes penetrating Carboniferous rocks. Relatively uniform log responses assist in delimiting the stratigraphical boundaries of volcanic rocks downhole. The Larne No. 2 Borehole (Penn *et al.* 1983) penetrated volcanic rocks of Lower Permian age and intrusive rocks of Tertiary age in a largely uncored borehole. Eight lithological subdivisions of the Lower Permian volcanic rocks were recognized on the basis of sample material and wireline log analysis.

Lower Palaeozoic log signatures

Thick Cambrian and Tremadoc shale sequences occur extensively in the subsurface of central and southern England and have been proved notably in boreholes at Cooles Farm and Shrewton (Whittaker 1980a). They comprise finely laminated siltstones, with mudstone and sandstone partings, and are remarkably uniform over considerable thicknesses. This is reflected in their log signature, which is characterized by a small range of variation in natural radioactivity or interval transit time (Fig. 13). Detailed log correlation of such successions may prove possible but insufficient drilling and logging has been carried out and the necessary initial palaeontological control is not to hand.

Ordovician and Silurian rocks have been proved beneath Upper Palaeozoic rocks in deep boreholes drilled in the Welsh Borderlands and the South Wales Coalfield, e.g. Collington, Senghenydd and Maesteg. Log analysis suggests that many of the major formations identified in the nearby outcrops of the Welsh Borderlands, e.g. Coalbrookedale Formation (Wenlock Shale), Much Wenlock Limestone (Wenlock Limestone), can be broadly recognized in the subsurface (Fig. 14), but detailed log correlation cannot yet be attempted. Scattered provings of established or probable Ordovician and Silurian rocks in the subsurface of the more easterly parts of southern England at present provide little information to aid detailed log correlation.

Devonian and Old Red Sandstone log signatures

Devonian rocks, in places of great thickness, are widespread in Britain. They belong mainly to the Old Red Sandstone facies (which may include rocks of late Silurian and early Carboniferous age) and are predominantly fluvial and alluvial fan deposits. The main use of logs in the study of these rocks is in broad lithological subdivision and in sedimentology. Detailed correlation could prove possible in the lacustrine Middle Old Red Sandstone of north-east Scotland and the marine Devonian strata of southern Britain, but to the authors' knowledge insufficient data exist to test this suggestion.

In southern Britain numerous deep boreholes cut Devonian rocks but, with a few important exceptions, the extent of penetration has been small. On the London Platform and westwards into the South Wales coalfield the Devonian rocks largely belong to the Lower Old Red Sandstone, of fluvial origin. Rocks of similar facies also occur in the Variscan foldbelt. Typical log signatures are given in Fig. 15, which shows the Lower Old Red Sandstone proved in the Maesteg Borehole and the ?Middle Devonian

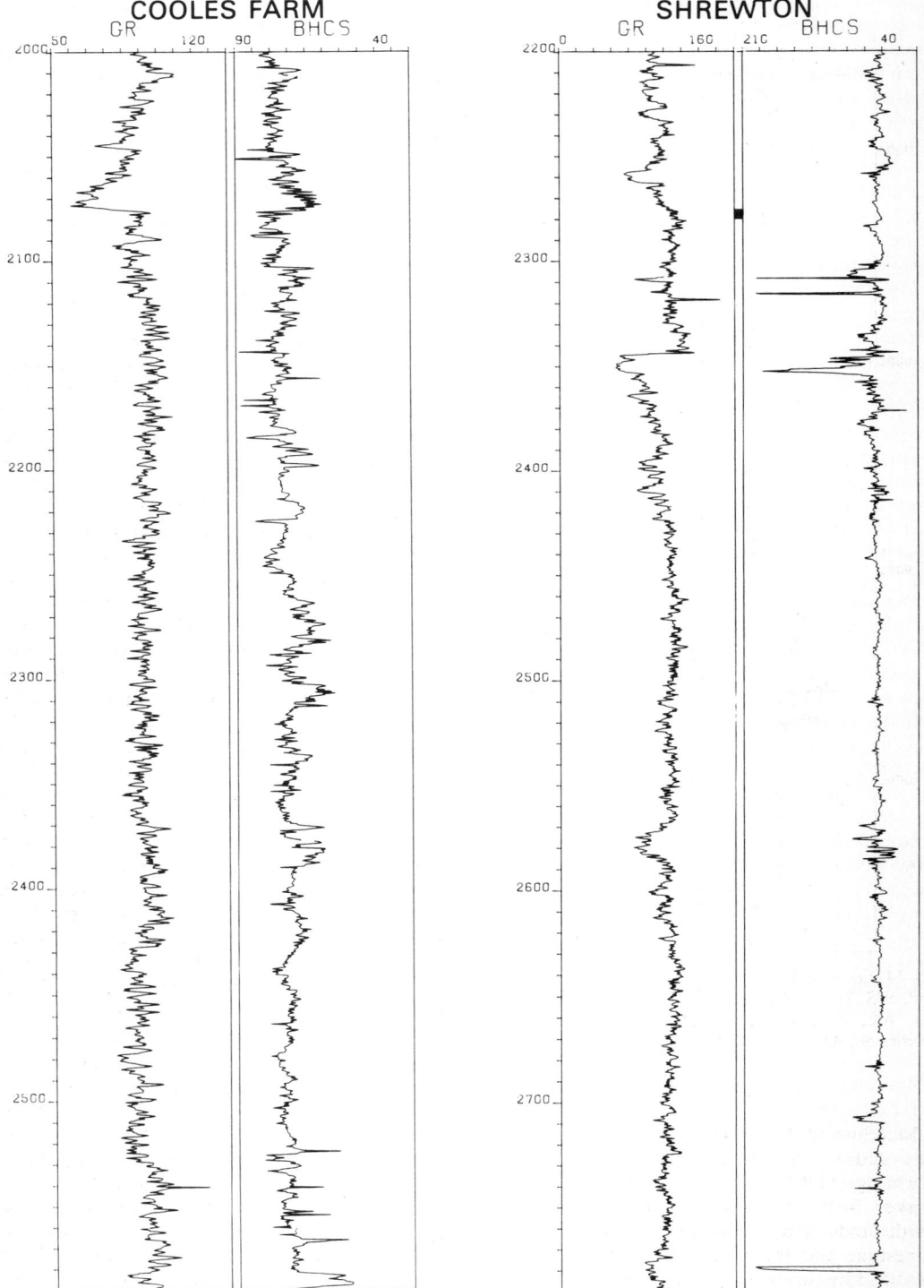

FIG. 13. Cambro-Ordovician, including Tremadoc (undifferentiated) log signatures: southern England.

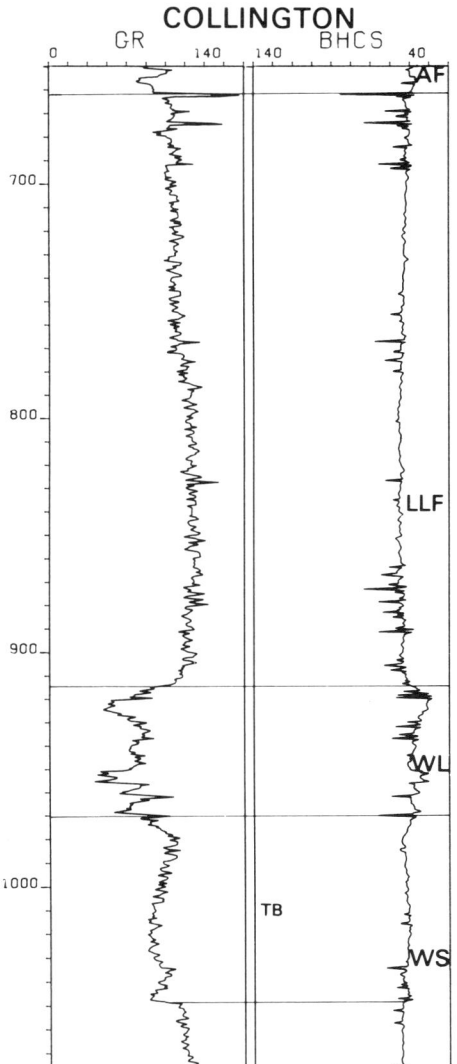

COLLINGTON

FIG. 14. Lower Palaeozoic log signatures: Welsh Borderland. WS, Wenlock Shale; TB, Tickwood Beds equivalent; WL, Wenlock Limestone; LLF, Lower Ludlow Formation; AF, Aymestry Formation.

at Marchwood (Southampton). The former example is additionally notable since, as pointed out by Tunbridge (1983), two important stratigraphical marker beds known at outcrop in the Welsh Borderlands and elsewhere, the *'Psammosteus'* Limestone and the Townsend Tuff Bed, can be identified by careful examination of the geophysical logs. The highly serrated log traces given by the Old Red Sandstone result from the regular repeti-

tion of sandstone, siltstone and mudstone. The sandstones are rarely more than 5 m thick and the log traces suggest that many of these are sharply based and form fining-upwards cycles.

Lower Carboniferous (Dinantian) log signatures

Dinantian rocks occur commonly throughout Britain, except in northern Scotland. They exhibit, both regionally and locally, a wide range of facies and thickness, which has been fully documented elsewhere (e.g. George *et al.* 1976). The main palaeogeographical/facies divisions recognized are the Culm facies of the Variscan foldbelt; the Carboniferous Limestone of South Wales, Bristol-Mendips and eastward subsurface extensions, North Wales and north Midlands (Derbyshire and westerly and easterly subsurface extensions), and the Yorkshire Dales–southern Lake District area; the mudstone/limestone infill of the Craven and satellite basins; and the largely fluvio-deltaic facies of northern England and Scotland.

The use of geophysical logs in correlation is more advanced in the Carboniferous Limestone facies, especially in the area to the east of the outcrops of the Derbyshire Dome (Nottinghamshire, Lincolnshire and adjacent areas) where Lower Carboniferous rocks have been penetrated in numerous hydrocarbon and coal exploration boreholes. With the wealth of detailed knowledge of the Derbyshire outcrops and an increasing amount of subsurface palaeontological control as guides, log correlation is proving a useful stratigraphical tool. Unfortunately many of the data remain confidential and only an outline account can be presented here. The fully cored BGS Eyam Borehole (Fig. 16) has proved the greatest thickness of Dinantian rocks in this area and provides a standard succession. Stage boundaries in the borehole have been established broadly by palaeontological analysis; the positions of these as indicated in Fig. 16 are taken at the nearest appropriate geophysical marker or lithological change consistent with the fossil evidence. For the most part the gamma ray log shows very low radioactivity levels. Correlatable features have proved to be shale partings with high gamma levels, typically in rocks of Asbian and Brigantian age, and more general minor shifts in gamma ray values, notably in beds of Chadian age. An argillaceous unit, at or near the top of the Arundian (the 'Arundian Shale Marker') shows higher than usual gamma values. It is a widely recognized marker on logs in this area but is not well developed at Eyam.

Elsewhere, even though extended drilling through Carboniferous Limestone with adequate geophysical logging and palaeontological control is

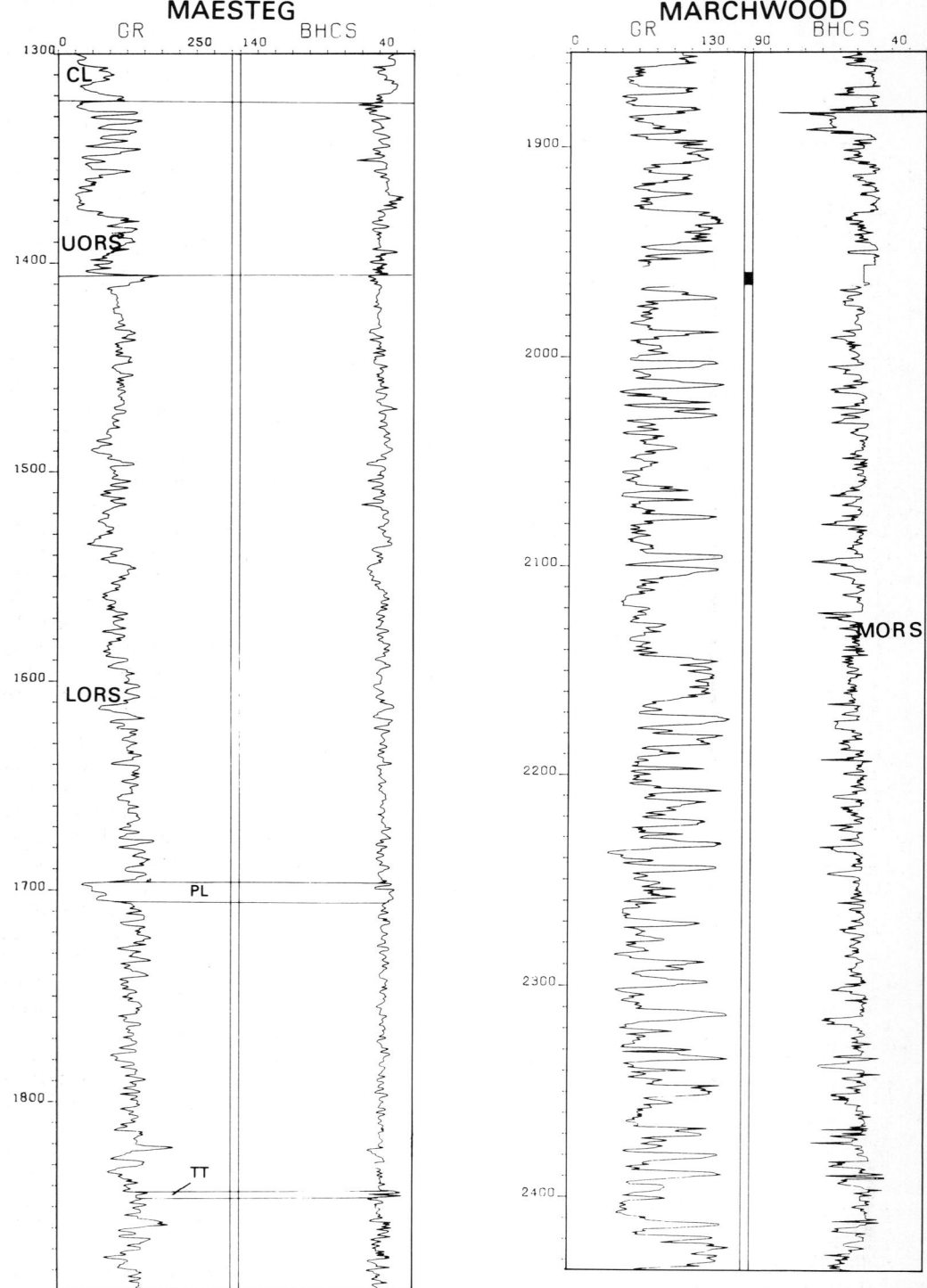

Fig. 15. Devonian (Old Red Sandstone) log signatures: Wales and southern England. LORS, Lower Old Red Sandstone; TT, Townsend Tuff Bed; PL, *'Psammosteus'* Limestone; MORS, Middle Old Red Sandstone; UORS, Upper Old Red Sandstone; CL, Carboniferous Limestone.

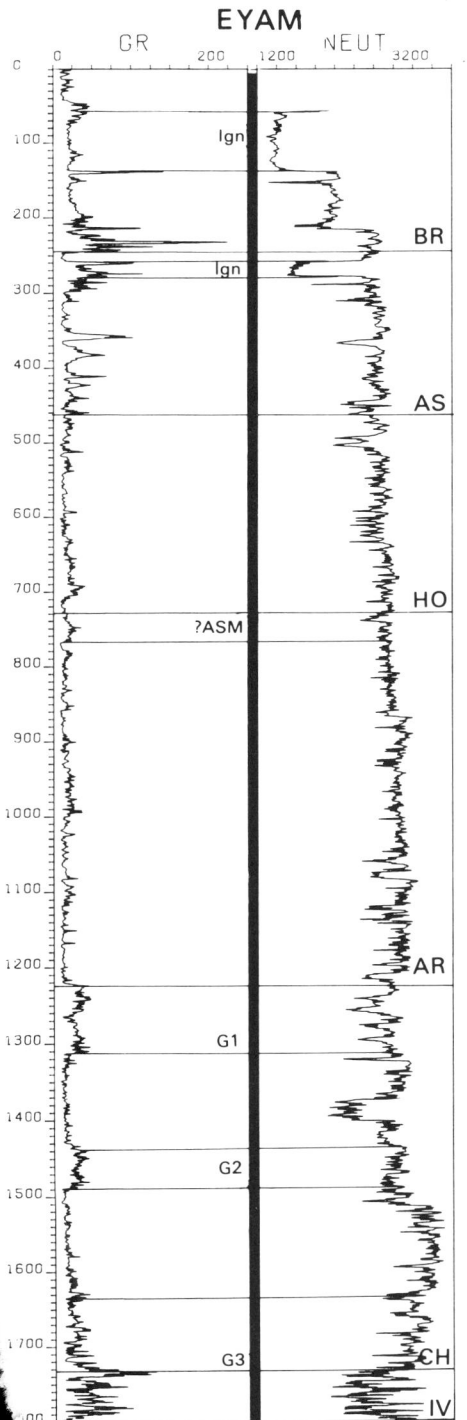

EYAM

16. Dinantian log signatures: Derbyshire. ORD, ~vician; IV, Ivorian; CH, Chadian; G1, G2, G3 are ~nal gamma ray markers; AR, Arundian; ?ASM, ~ndian Shale Marker'; HO, Holkerian; AS, Asbian; ~neous rocks; BR, Brigantian.

limited, there is evidence of the important role that geophysical logs can play in borehole correlation. The BGS cored boreholes at Warlingham (Surrey) and Cannington Park (Somerset) both proved rocks of Tournaisian age and there are many points of similarity on their respective gamma ray log traces (Fig. 17). More remarkable are the similarities both show with the log from an open-hole borehole through the whole Carboniferous Limestone at Maesteg (South Wales) (Fig. 17). In addition the upper part of the Maesteg Dinantian is characterized, as is that at Eyam, by mudstone partings with high gamma ray values. Using such criteria, in the absence of palaeontological control, the suggested positions of the major stage boundaries in the Maesteg Borehole are shown in Fig. 17. There are few detailed similarities between the log traces observed in southern Britain with those in Derbyshire (Fig. 16), apart from the responses from rocks of Asbian and Brigantian age; for example, the thin Chadian sequences of Somerset and South Wales contrast with the much thicker succession proved at Eyam and elsewhere in the Nottinghamshire–Lincolnshire subsurface. Such observations give some indications of the likely limitations on the future use of log correlation in the study of the Carboniferous Limestone. In addition, in those areas of incomplete or seriously interrupted sedimentation, log correlation without palaeontological control must proceed with the utmost caution.

The Dinantian rocks of northern England and Scotland contain a wide range of lithologies including limestone, ironstone, mudstone, siltstone, sandstone, seatearths and coal. That these rock types commonly occur in repeated rhythmic sequence has been known since the early nineteenth century. Individual rhythmic units (cyclothems) range in thickness from less than 1 m to up to 50 m, and may be local or regionally widespread in extent. The lithological division, and to some extent in practice the chronostratigraphic subdivision, of these rocks depends in large measure on the recognition of key marker horizons as the major facies changes are generally ill-defined and diachronous. In the past few decades many important stratigraphical boreholes have been cored with, in a few cases, geophysical logs. For the most part the logs are unpublished and in some cases remain confidential. Two examples of such boreholes (Fig. 18) from the Upper Border and Liddesdale groups of Northumberland, show characteristic log responses and confirm that the main limestones, sandstones and coals in this type of succession can be identified from geophysical logs. Thus one can with confidence identify the thicker marker beds in

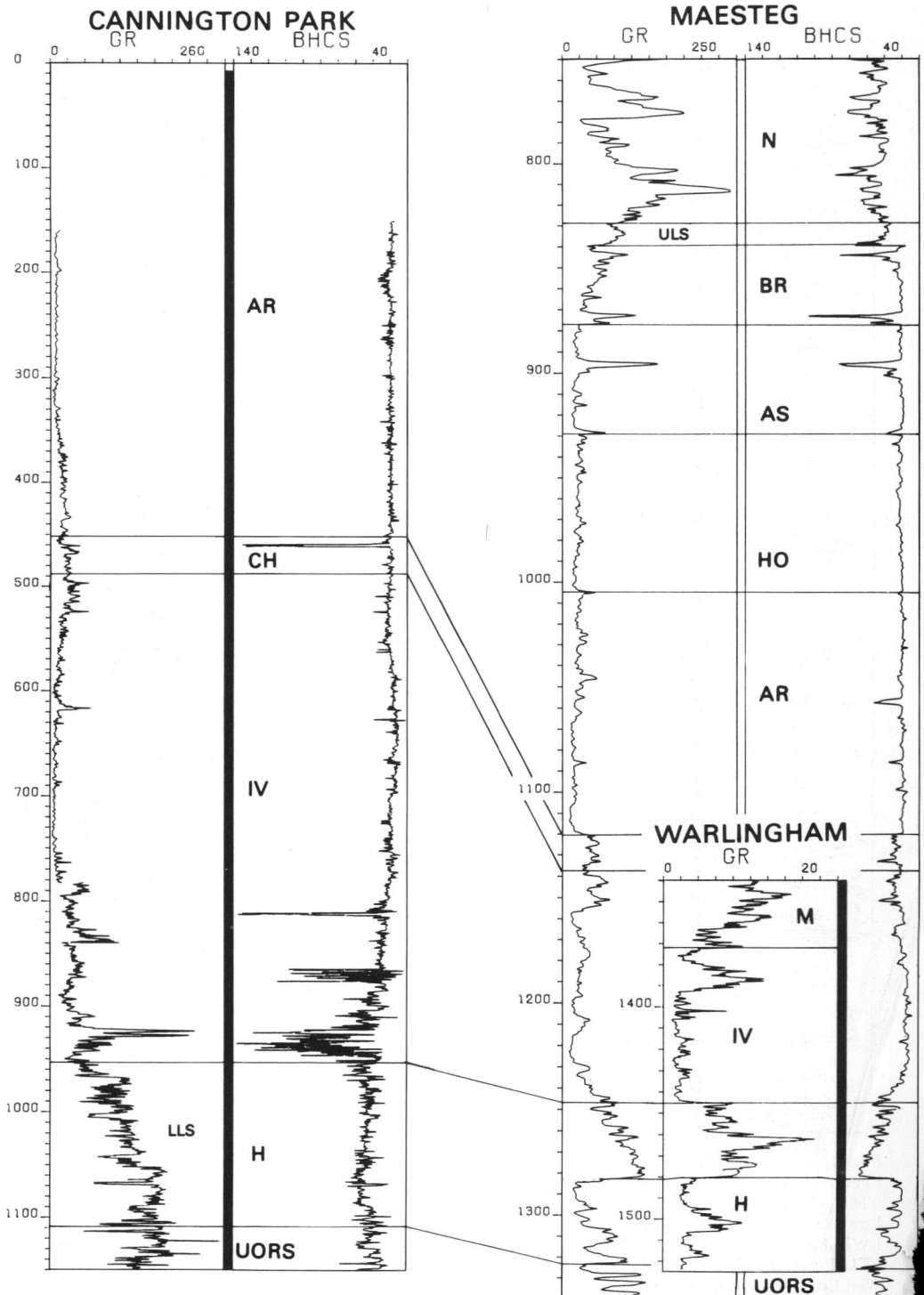

FIG. 17. Dinantian log signatures: south of St George's Land. UORS, Upper Old Red Sandstone; LLS, Lower Limestone Shale; H, Hastarian; IV, Ivorian; CH, Chadian; AR, Arundian; HO, Holkerian; AS, Asbian; BR, Brigantian; ULS, Upper Limestone Shale; N, Namurian; M, Mesozoic.

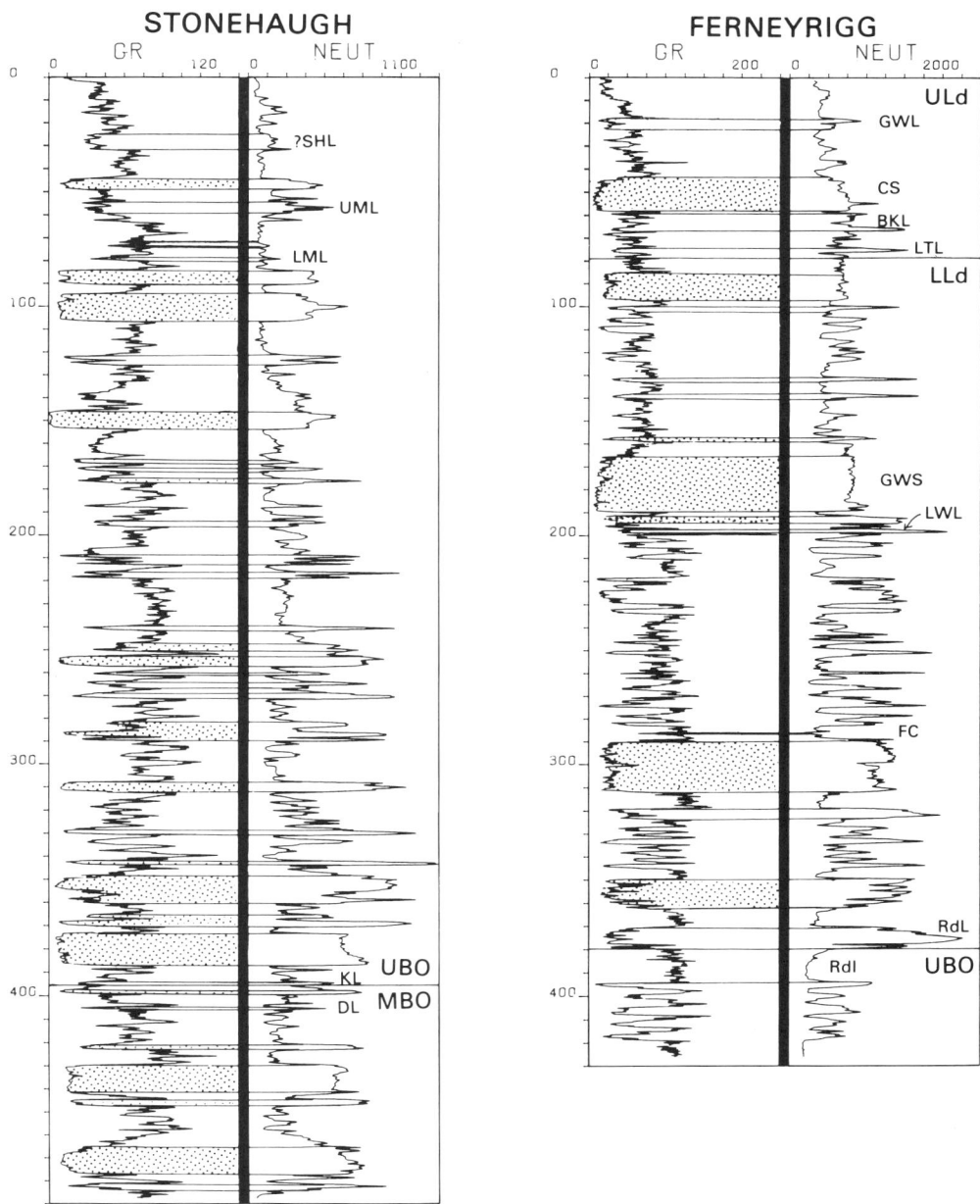

Fig. 18. Dinantian log signatures: Northumberland. MBo, Middle Border Group; DL, Desoglin Limestone; UBo, Upper Border Group; KL, Kingbridge Limestone; LML, Lower Millerhill Limestone; UML, Upper Millerhill Limestone; ?SHL, ?Spy Hole Limestone; RdI, Redesdale Ironstone Shale; LLd, Lower Liddesdale Group; RdL, Redesdale Limestone; FL, Fourlaws Limestone; FC, Fourlaws Coal; LWL, Ladies Wood Limestone; GWS, Great Wanney Crag Sandstone; ULd, Upper Liddesdale Group; LTL, Lower Tipalt Limestone; BKL, Bankhouses Limestone; CS, Camphill Sandstone; GWL, Greengate Well Limestone. Main sandstones stippled; all logs in counts/s.

commercial open-hole borings from analysis of the geophysical logs and cutting samples. One example is provided by the Harton Borehole (geophysical logs not yet released for publication) where detailed correlation has been achieved with successions established at outcrop 50 to 70 km to the west and north-west (Ridd *et al.* 1970, Frost & Holliday 1980). However, it is likely that thinner and lithologically less distinct marker bands such as are more typical of the early and mid-Dinantian, will be more difficult to recognize without core. Many of the key marker beds in the Border Group of the Northumbrian succession are fossiliferous shales or thin impure limestones (both sandy and argillaceous) which can be recognized on geophysical logs only with difficulty. Given the great lateral and vertical variation in these rocks, the presence of widely recognized regional geophysical markers is unlikely to be common. One such marker could prove to be the very high gamma ray values observed immediately below the Ladies Wood (= Denton Mill) Limestone in the Ferneyrigg Borehole (Fig. 18), which have been observed also at the same horizon in at least one commercial borehole. Scottish Dinantian successions have similar log characteristics to their northern England counterparts, but the limestones are generally thinner and less common, and volcanic rocks form a significant part of the sequence. Published geophysical logs are available for the BGS Spilmersford Borehole (Davies 1974, Allsop 1974).

There are few deep boreholes which prove significant thicknesses of Dinantian rocks in the Craven Basin or in rocks of basinal facies around Derbyshire. Of these, many were not logged geophysically or are commercial in confidence. A gamma ray log of the BGS Duffield Borehole, which proved late Dinantian mudstones with limestone and sandstone turbidites and volcanics, has been published (Aitkenhead 1977), but the lack of detail and character on the log in this part of the proven succession suggests that the recording may have been faulty.

Upper Carboniferous (Silesian) log signatures

Upper Carboniferous (Silesian) rocks are the oldest in Britain for which detailed log correlation is both well established and advanced. This stems from their economic significance and the resultant high level of exploration activity to investigate their nature and extent. Most exploration boreholes of the NCB drilled in the last decade have extensive log suites, particularly those drilled to prove new coalfields or concealed extensions of established areas of working. The geophysical logs are used for a number of purposes including stratigraphical correlation and coal seam thickness determinations in both the open-hole and cored sections of the hole. However, in the area to the east of the Yorkshire–Nottinghamshire coalfield there are in addition numerous hydrocarbon wells, with detailed geophysical log suites, which unlike most of the NCB boreholes prove the Namurian rocks beneath the Westphalian Coal Measures. Thus there is a wealth of data here which make the Silesian rocks among the best known of that age in Britain, in an area where borehole correlation by use of geophysical logs is a standard every-day operation. As most of the data remain confidential the following account, which concentrates on this better known area, is of necessity somewhat generalized.

Analyses of Silesian sediments have suggested that the dominant origin of radioactivity is the isotope ^{40}K. The uranium and thorium content is for the most part low and finely disseminated. The main exceptions are to be found in certain marine shales which commonly exhibit higher than average levels of radioactivity resulting from the presence of uranium absorbed in phosphatic fish remains and collophane and with certain types of carbonaceous and bituminous matter (Ponsford 1955, Ramsbottom *et al.* 1962, Knowles 1964). The log traces shown in Figs 19 to 22 are typical of Silesian rocks in many parts of the country. Siltstone is generally the most abundant rock type closely followed by sandstone and mudstone; coal seams form only a small proportion of the whole. The rapid alternation of these lithologies produces log traces of considerable character from which the main rock types can be identified. Sandstones are among the most noticeable feature of the logs. Unfortunately, they are known to be impersistent and variable, even in the Millstone Grit, and, except locally or in restricted parts of the succession, they are of limited value in correlation. Coal seams are readily identified and can be used as geophysical log markers. Many are laterally persistent, but all are subject to local thinning and splitting or absence due to non-deposition or contemporaneous erosion ('wash-outs'). Mudstones, with higher than normal gamma values, are also potential log markers, particularly the more persistent marine bands, which form the basis of both the lithological and chronostratigraphical subdivision of the Silesian rocks of Britain. Their identification is a prime target for geophysical log correlation which can be achieved with varying degrees of success. The highest gamma ray values recorded generally correspond to the marine bands of the Millstone Grit and Coal Measures, although

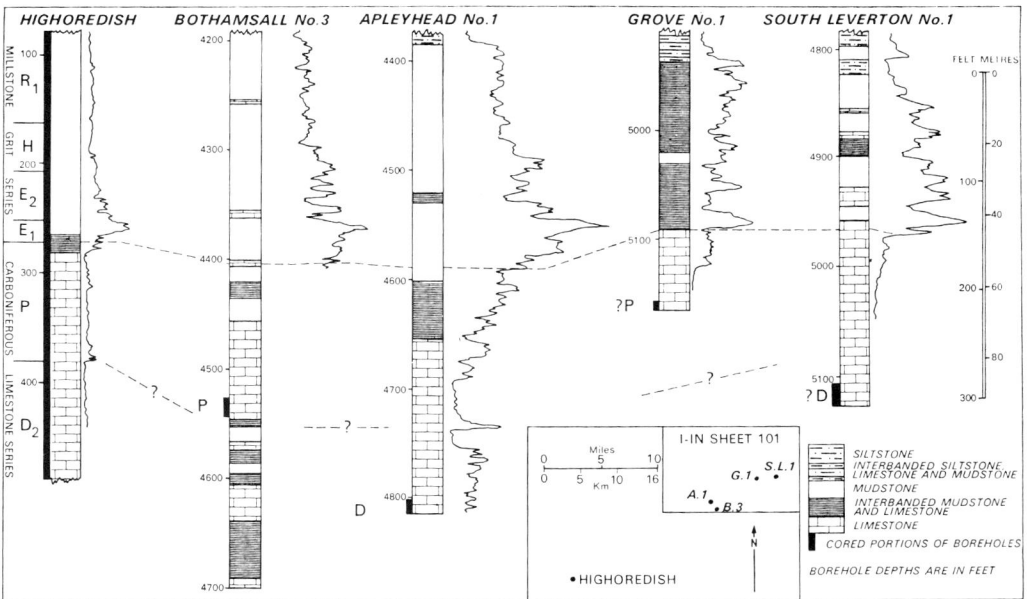

FIG. 19. Gamma ray logs of late Dinantian and early Silesian rocks in some boreholes in Derbyshire and Nottinghamshire (from Smith *et al.* 1973).

this is not the case for all such bands and individual marine beds give different natural radioactivity readings from place to place. In practice detailed borehole correlation is achieved by using all available evidence from logs or cuttings and matching this with data from the nearest cored succession. Even where no cored borehole occurs nearby, a fairly full and broadly reliable correlation can be achieved from the study of the geophysical logs, particularly the gamma log.

In the subsurface of Nottinghamshire and Lincolnshire, there is commonly an unconformity between Silesian (Millstone Grit or Coal Measures) and Dinantian (Carboniferous Limestone) rocks, readily observed at the sharp boundary where mudstones and siltstones with sandstones rest on limestone. In other parts of the area there is no break in sedimentation and the Silesian/Dinantian boundary occurs at the *Cravenoceras leion* Marine Band within a dominantly argillaceous sequence. Comparison of open-hole boreholes with those cored and geophysically logged (Ramsbottom *et al.* 1962, Smith *et al.* 1973, fig. 4) enables the position of the *C. leion* band to be selected in the former with some accuracy (Fig. 19). Similar log patterns in shales spanning the Dinantian/Silesian boundary have been observed in deep boreholes to the west in Lancashire and Chesire and, apparently, as far

afield as the Maesteg Borehole in South Wales (Fig. 17). In addition to the logs shown here, gamma ray logs, at a much reduced scale, of Namurian rocks in many of the oil boreholes of the East Midlands have been published by Downing & Howitt (1969). A common feature of these logs is the relatively high level of radioactivity in beds of Pendleian to Alportian age (Ramsbottom *et al.* 1962) (Figs 17 and 19). Within this highly radioactive zone, individual peaks, some of which can be related to goniatite-bearing bands, can be correlated at least locally. The high level of uranium in these beds perhaps results from their deposition in relatively deep water away from deltaic and terrestrial influences. The top of this high gamma zone is diachronous due to the incoming, mainly from the north, of increasing volumes of turbiditic and deltaic sediments. However, early Pendleian sediments in the Craven Basin, as proved around Burnley and Harrogate, are characterized by similar high gamma values. Farther north in north-east England, marine shales at comparable levels associated with the Great and Little Limestones, also have higher gamma values than similar shales at other levels in a number of boreholes. As far south as South Wales, early Pendleian shales are also characterized by high gamma ray values (Fig. 17).

In the East Midlands, strata of Kinderscoutian

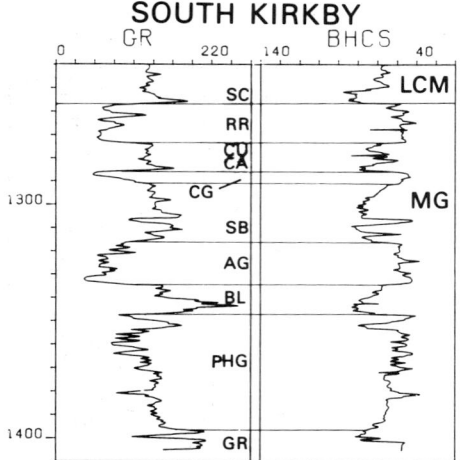

FIG. 20. Late Namurian log signatures: Yorkshire. MG, Millstone Grit; GR, *Bilinguites gracile* Marine Band; PHG, Pule Hill Grit; BL, *B. bilingue* (late form) Marine Band; AG, Ashover Grit; SB, *B. superbilingue* Marine Band; CG, Chatsworth Grit; CA, *Gastrioceras cancellatum* Marine Band; CU, *G. cumbriense* Marine Band; RR, Rough Rock; LCM, Lower Coal Measures; SC, Subcrenatum Marine Band.

to early Westphalian A age typically show serrated log traces with the highest gamma values generally less than in Pendleian beds (Fig. 20). The major grits show prominently as gamma lows, and available core material shows that the higher gamma values shales commonly correlate with major, goniatite-bearing marine bands. The identification from geophysical logs of the main goniatite bands and the well-developed, named grits allow for refined correlation locally. However, as individual beds are not present everywhere and vary laterally in thickness and nature (including level of radioactivity), great caution must be shown in correlation by such means, especially where the borehole concerned is remote (i.e. more than a few kilometres) from others with palaeontological control. Probably the most easily identified horizon is the Listeri Marine Band, which commonly shows high radioactivity, up to twice the counts of many other shales and comparable to Pendleian shales. Using this marker Howitt (Brunstrom 1963, Downing & Howitt 1969; see also Edwards 1967) was able to show that the correlation of the Westphalian/Namurian boundary in many boreholes by some of the early workers in the East Midlands has been incorrect. Occurrences of *Gastrioceras listeri* and *G. subcrenatum* in numerous cores since have shown the validity of Howitt's work. However, more recent drilling also has

proved an unfortunately high number of boreholes where the Listeri Marine Band is not specially marked by greater than normal radioactivity and thus is not prominent on those gamma ray logs.

In other areas peripheral to the southern and central Pennines similar log characteristics have been observed in Kinderscoutian to Lower Westphalian A strata. In boreholes situated close to outcrop, where there are well-established successions, detailed correlation of grits and marine bands can be readily achieved. However, in localities more remote from the outcrop, e.g. Cheshire Basin, such detailed work cannot be attempted yet without palaeontological control. Available data suggest that high radioactivity levels at the horizon of the Listeri Marine Band are not found generally away from the East Midlands. Thus this band, and more importantly the horizon of *G. subcrenatum*, cannot be fixed with certainty in many cases by examination of geophysical logs alone.

In the northern Pennines, north-east England and Scotland, the Namurian rocks are only partially of Millstone Grit facies showing instead many similarities to the underlying Dinantian strata. In the Pendleian to early Kinderscoutian the major marine beds are limestones rather than goniatite-bearing black shales. These can be recognized readily from geophysical logs and correlated with known successions. Much greater difficulty is encountered with Kinderscoutian to early Westphalian A beds; the detailed faunal succession in these beds is poorly known at outcrop and in cored boreholes because the goniatite bands of farther south are here represented by *Lingula* bands or sandy/limey shell beds, the precise correlation of which is uncertain, and which are not everywhere present due to non-deposition or contemporaneous erosion.

The main productive Lower and Middle Coal Measures (Westphalian A–B) are lithologically similar throughout the major coalfields of Britain; this similarity is shared by their geophysical log traces. These beds, in common with the underlying Namurian rocks, have serrated log traces brought about by the alternation of sandstones and shales. However, the Westphalian rocks have a more pronounced spiky pattern, particularly on gamma ray and density logs, resulting from the presence of numerous thick coal seams (Fig. 21). By way of contrast with the spiky, low gamma levels from the coal seams and sandstones, the argillaceous beds have a generally moderate and uniformly consistent level of radioactivity. The Vanderbeckei Marine Band only rarely exceeds the general level of shale radioactivity and as a geophysical log marker is of limited value on its own. The numerous marine

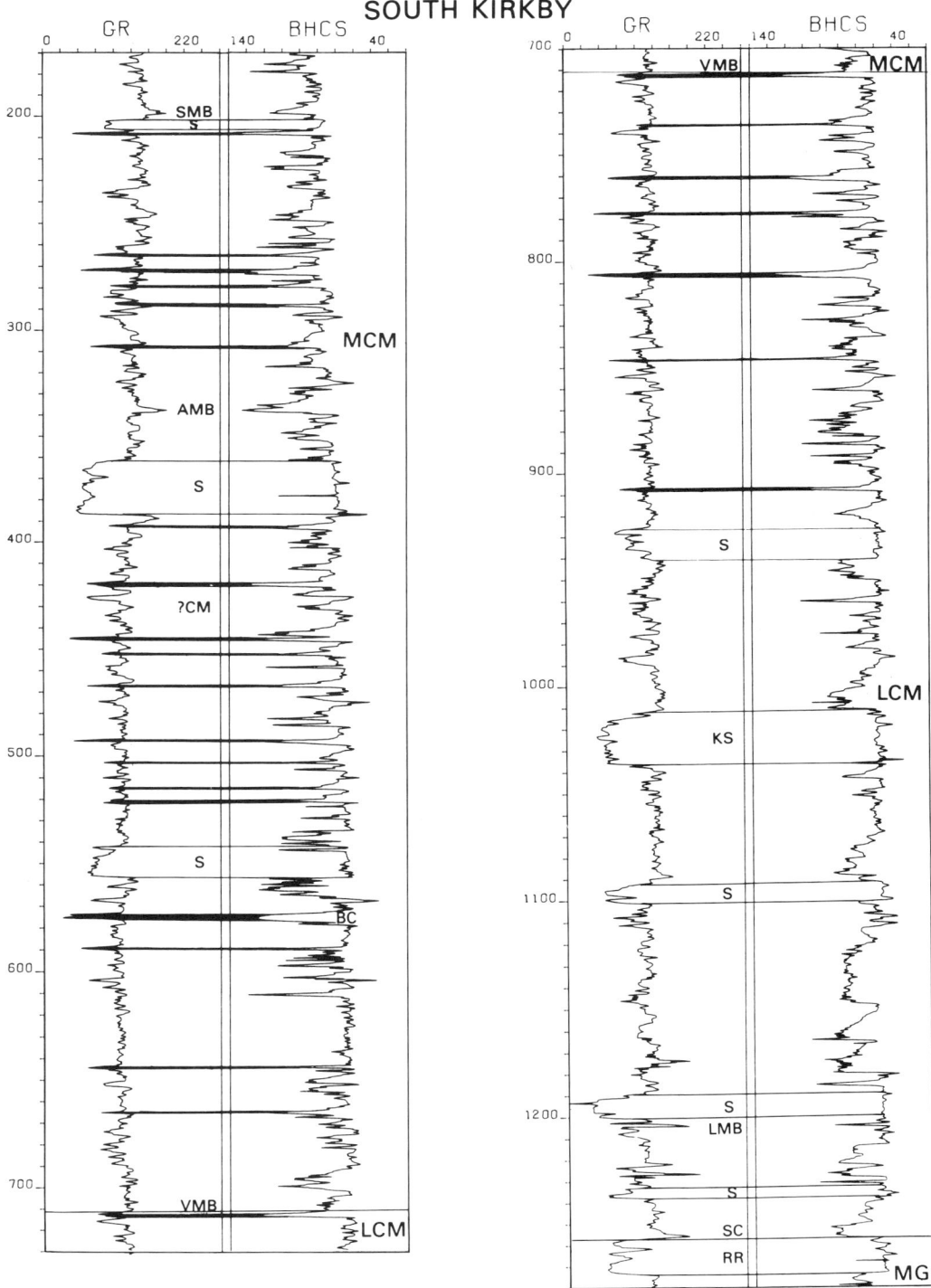

FIG. 21. Westphalian log signatures: Yorkshire. MG, Millstone Grit; RR, Rough Rock; LCM, Lower Coal measures; SC, Subcrenatum Marine Band; S, sandstone; LMB, Listeri Marine Band; KS, Kilburn Sandstone; MCM, Middle Coal Measures; VMB, Vanderbeckei Marine Band; BC, Barnsley Coal; ?CM, ?Clowne Marker; AMB, Aegiranum Marine Band; SMB, Shafton Marine Band.

MILTON GREEN

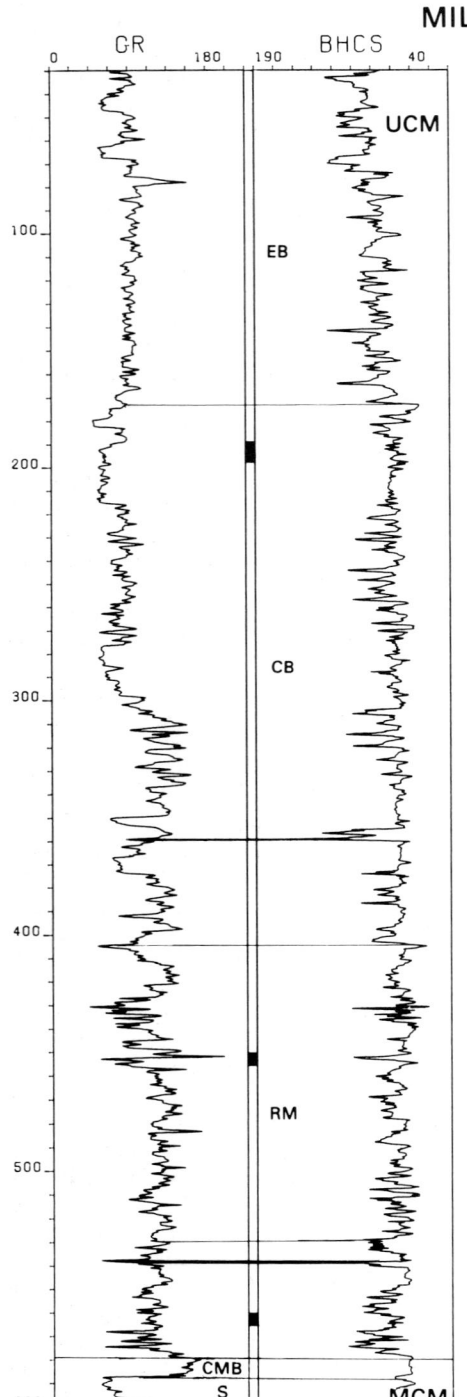

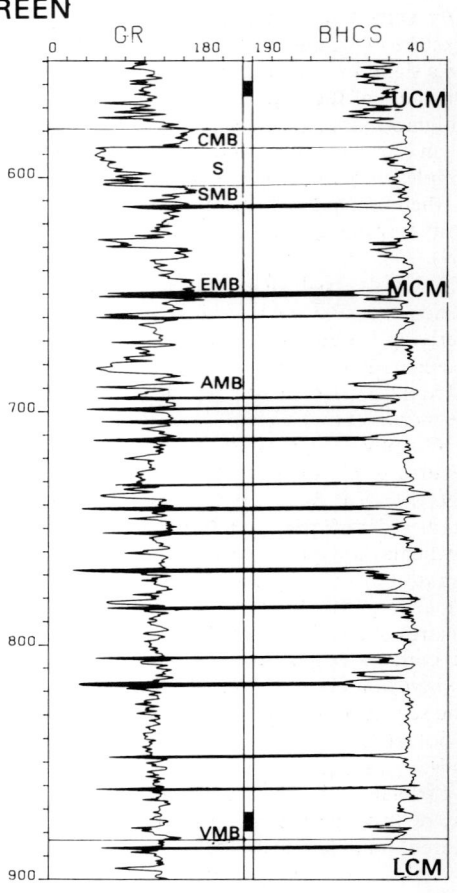

FIG. 22. Westphalian log signatures: Cheshire. LCM, Lower Coal Measures; MCM, Middle Coal Measures; VMB, Vanderbeckei Marine Band; AMB, Aegiranum Marine Band; EMB, Edmondia Marine Band; SMB, Shafton Marine Band; S, sandstone; CMB, Cambriense Marine Band; UCM, Upper Coal Measures; RM, Ruabon Marl; CB, Coed-yr-Allt Beds; EB, Erbistock Beds.

bands that occur towards the top of Westphalian B and the base of Westphalian C similarly only locally exceed the general average value. In the Yorkshire–Nottinghamshire coalfield, boreholes show a slight increase in general shale radioactivity in this part of the sequence (also see Fig. 22), but the highest gamma values locally correspond to a level in the seatearth beneath the Clowne Coal. Examples of gamma ray logs from this coalfield and their correlation have been illustrated by Howitt & Brunstom (1966) and Downing & Howitt (1969).

The Upper Coal Measures (Westphalian C–D) contain fewer and generally thinner coals than the underlying Productive Coal Measures and this is reflected in their log traces. Figure 22 shows Upper Coal Measures from Cheshire and is typical also of these measures in the West Midlands and Lancashire. Geophysical logs from the Upper Coal Measures of Oxfordshire are illustrated in Poole (1969, 1977, 1978). These measures commonly contain red beds (Etruria Marl, Keele Beds and equivalents) and in some areas thick sandstones (Pennant Group of South Wales and Arenaceous Coal Group of Oxfordshire). The Upper Coal Measures of the Yorkshire–Nottinghamshire coalfield contain a number of faunal markers but elsewhere faunal control is poor. Coring in this part of the sequence is generally limited and its economic potential to the coal and hydrocarbon industries is low. For these various reasons geophysical log correlation is little advanced, except where coals of this age are being actively exploited or explored, but a large volume of (commercial) data is available and offers the hope of future progress.

Permian log signatures

Permian rocks in western Britain are largely non-marine red beds, consisting of mudstones, siltstones, sandstones and conglomerates, with little faunal or floral control of their precise age. Geophysical logs have proved very useful in borehole correlation of these rocks, notably in the north-west. A few provings of assumed Permian rocks have been made in south-west England, mostly in confidential boreholes. Figure 23 illustrates an example from the argillaceous Aylesbeare Group from the Winterborne Kingston Borehole, but this may not be typical of the area; comparison with outcrops and other boreholes suggests that these strata are variable both in thickness and lithology, with sandstones and breccias occurring elsewhere. On the other hand the clean, relatively well sorted Bridgnorth Sandstones (Lower Mottled Sandstone), exhibiting large-scale cross-stratifica-

tion, of the Kempsey Borehole (Fig. 23) seem typical of many boreholes and outcrops in the Worcester/Severn Basin area extending into the southern Cheshire Basin. Farther north the equivalent Kinnerton Sandstone contains an increasingly important intercalation of shales, some containing marine fossils, and evaporites (the Manchester Marl and/or the St Bees Evaporites and Shales) thought to be of late Permian age. Examples of geophysical logs and their correlation from Cheshire to the Fylde and into the Irish Sea have been given by Colter (1978) (see also Burgess & Holliday 1974, Penn *et al.* 1983). Regionally these Upper Permian sequences are very variable, both in lithology and thickness. Log analysis has assisted in the identification of these argillaceous and evaporite beds, but at present has proved mainly of local value in detailed correlation.

By way of contrast, geophysical logs have proved to be of immense value in correlating Permian sequences in the main area of British deposition of the southern North Sea Basin and landward extensions in eastern England from Norfolk to Northumberland.

Given the general lack of faunal control, lithological factors assume greater significance in correlation and, as the great majority of drilling is open-hole, geophysical logs are therefore a prime method of lithological determination. Indeed, given the varied evaporitic depositional environment it might be expected that log correlation would provide a more precise correlation than could be expected from faunal or floral analysis. Thousands of geophysically logged open-hole boreholes, with a limited number of cored controls, have been drilled in the course of hydrocarbon and coal exploration. Fortunately many hydrocarbon boreholes are released for public use and as a result a significant literature exists in which examples of log traces and their correlation are illustrated.

The Lower Permian (Rotliegendes) beds are generally thin onshore but thicken into the North Sea where they form the main reservoirs in the gas fields. Dominantly arenaceous they can also contain significant quantities of conglomerate and evaporitic siltstones and mudstones. The main uses of geophysical logs with these sediments fall not so much in the field of stratigraphical correlation but rather in the field of facies analysis discussed elsewhere in this report. The best reservoir rocks belong to an aeolian dune facies consisting of clean, well-sorted sandstones. Those sandstones and conglomerates of alluvial fan/sheet flood/wadi origin which are argillaceous and less well sorted, and in particular the argillaceous evaporitic interdune sabkha facies, make poor reservoirs. The differ-

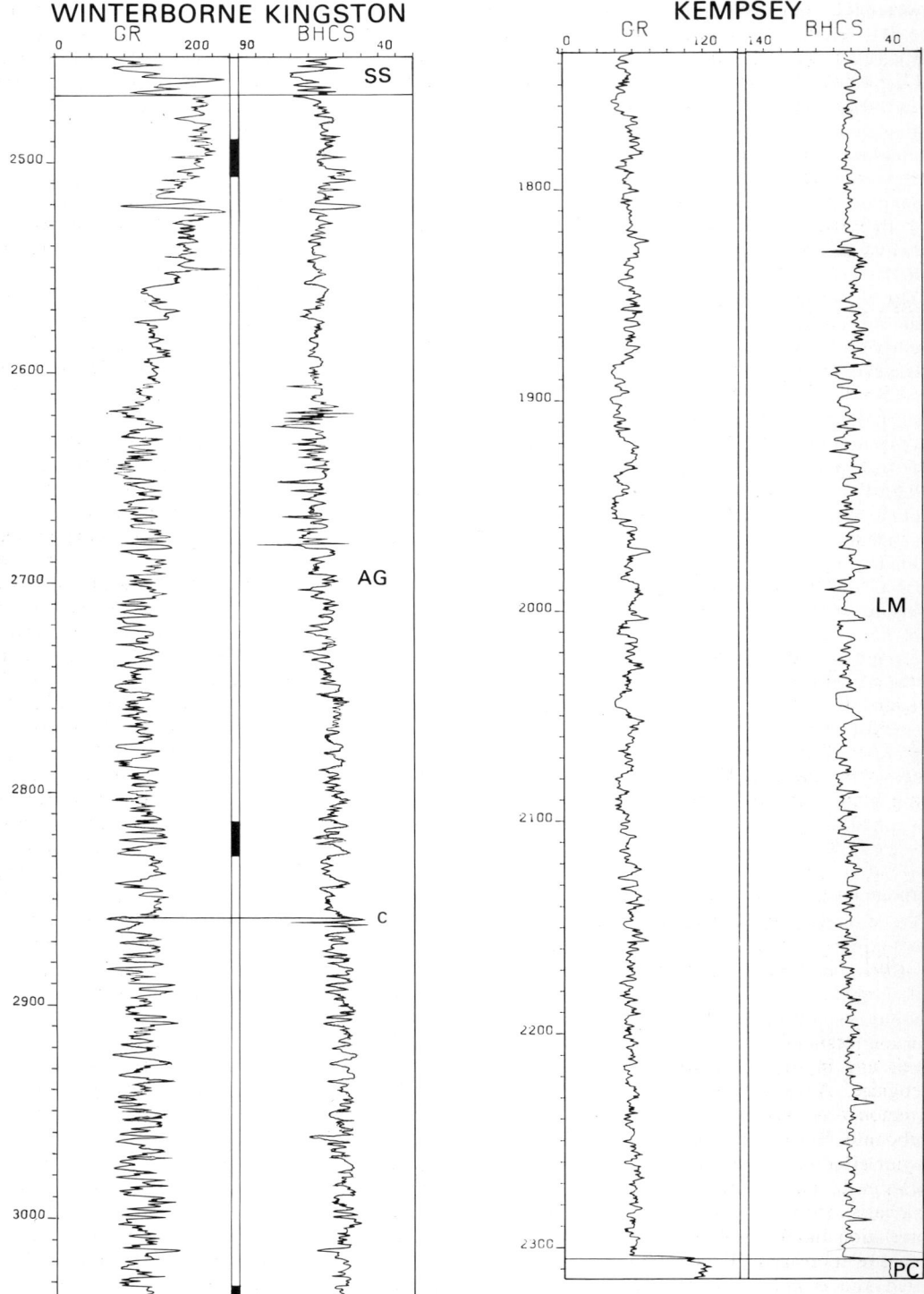

FIG. 23. Permian log signatures: Dorset and Worcester. PC, Precambrian; AG, Aylesbeare Group; C, colour change in cuttings; LM, Lower Mottled Sandstone; SS, Sherwood Sandstone Group.

Telephone

Date.. 7/9/92........

Name.. Shaun Oakey....

Address .. Home....

....

Phone No. ....
0980 630 382

Please call

Ivy PD1   Taken by .. Joanne

ences in lithology and physical properties between these facies often can be detected by examination of the common geophysical logs.

The Upper Permian or Zechstein rocks of eastern England and the southern North Sea are predominantly marine in origin or at least marine-influenced; continental red beds become more prominent around the basin margin and later extend across the whole basin and cover the earlier marine rocks. The marine rocks are dominantly carbonates (limestones and dolomites) and evaporites (anhydrite, halite and potash salts) which commonly do not feature prominently in cutting returns. With some important exceptions, noted below, most of these rock types have low radioactivity levels and initially the use of the gamma ray log is limited to distinguishing potash mineral, argillaceous interbeds and partings. Both carbonates and evaporites commonly contain small quantities of clay minerals and the resulting minor variations in the gamma ray trace have proved to be the basis for local detailed correlation of some significance. Perhaps more than for any other British system, the sonic log and more especially the density log are essential tools for sensitive lithological identification and separation and for stratigraphical correlation.

Two examples of gamma-sonic traces from Permian rocks from boreholes on the east coast of England are shown in Fig. 24. One, Ulceby Cross, illustrates an example close to the basin margin and the other, Barmston, towards the basin centre. These show the generally clear way in which the main lithological units can be distinguished. An account of the development of the British Zechstein Basin, illustrated with geophysical log traces and correlation, has been published by Taylor & Colter 1975 (see also Taylor 1981). The major Zechstein cycles Z_1 and Z_3 begin with sharp gamma ray highs resulting from thin basal argillaceous developments, the Marl Slate (Z_1) and Grauer Salzton (Z_3), but these may be overlapped towards the basin margin. Locally the Marl Slate rests directly on Carboniferous shales with similar radioactivity levels and in such instances may be difficult to recognize. A similar but smaller gamma peak commonly is observed at the base of the Z_2 carbonate. By such use of the gamma ray curve Taylor (1980) has recognized a number of minor cycles in Z_1 and the early part of Z_2 and used these as a basis for very detailed and refined lithological correlation and for establishing complex stratigraphical relationships. Work of this sort if substantiated and extended to other areas is likely to enhance greatly both knowledge and understanding of early Zechstein history and deposition. Such

relationships and correlations can be achieved in no other way than by use of geophysical logs. However, it should also be noted that Taylor's (1980) subcycles may not be recognizable everywhere in the basin and almost certainly not in a borehole considered in isolation from others. Indeed it is this part of the Zechstein succession that commonly provides the greatest difficulty to geologists in delineating the main lithological boundaries between formations that commonly consist of varying admixtures of calcite, dolomite and anhydrite and which exhibit considerable lateral thickness and lithological changes. Colter & Reed (1980) have used geophysical logs to correlate relatively thin cyclic units, originally identified in cores, in the Z_2 evaporites over a wide area of the Yorkshire coast and adjacent areas. Not only is such work of stratigraphical significance but if correct has important sedimentological implications indicating, as does Taylor's (1980) work on the Z_1 evaporites, a broadly foresetting mode of deposition and implying relatively deep water deposition. Smith & Crosby (1979) have published an account, with illustrative geophysical log correlation, of the third and fourth Zechstein cycles with particular reference to potash occurrences. Except in peripheral areas the top of the Zechstein is marked by a thin but prominent anhydrite band (Top Anhydrite) below the predominantly argillaceous Saliferous Marl (Bunter Shale) (Fig. 24). This fails towards the basin edge as the evaporites pass into red beds and first the 4th Cycle Upper Anhydrite, and then the Upper Magnesian Limestone, mark the top of the Zechstein. In the few metres above the Top Anhydrite, the Saliferous Marls are siltier than those overlying and this is reflected in the log traces which show a decrease in gamma values and in interval transit times. This siltier unit is known as the Bröckelschiefer and is widely recognized over the whole Zechstein Basin; its top can be recognized from geophysical logs in peripheral areas beyond the Top Anhydrite limit. Similar log pattern traces can be observed in some boreholes in north-west England at a similar level, e.g. between units xi and x of the Hilton Borehole in the Vale of Eden (Burgess & Holliday 1974), and this may prove to have wider stratigraphical significance.

While there are still some problems in correlation, especially in the delineation of formation boundaries, it is clear that Zechstein sequences can be correlated all the way from eastern England to Poland and that many individual named beds are continuous for that great distance. Perhaps in no other major group of rocks has geophysical log correlation shown its value over such a wide

ULCEBY CROSS

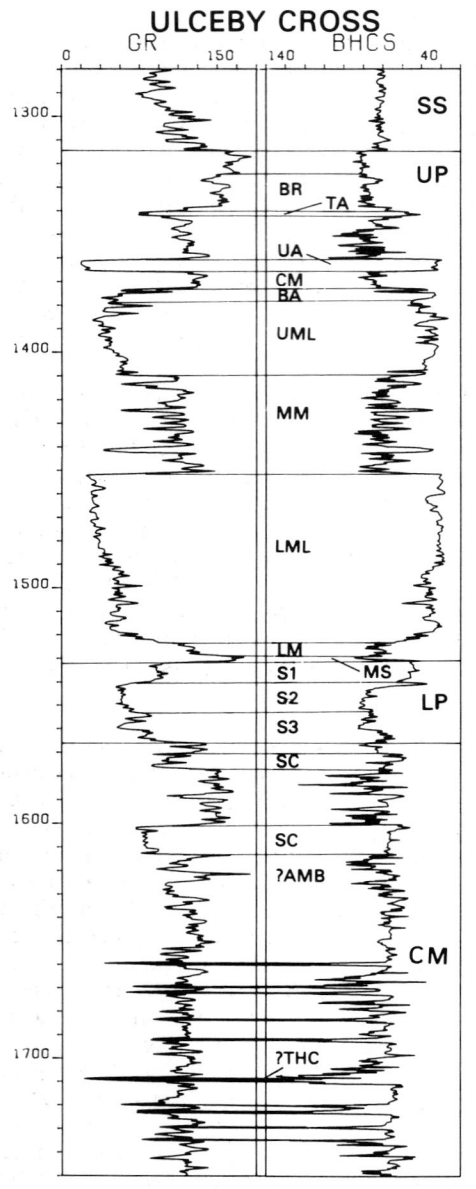

BARMSTON

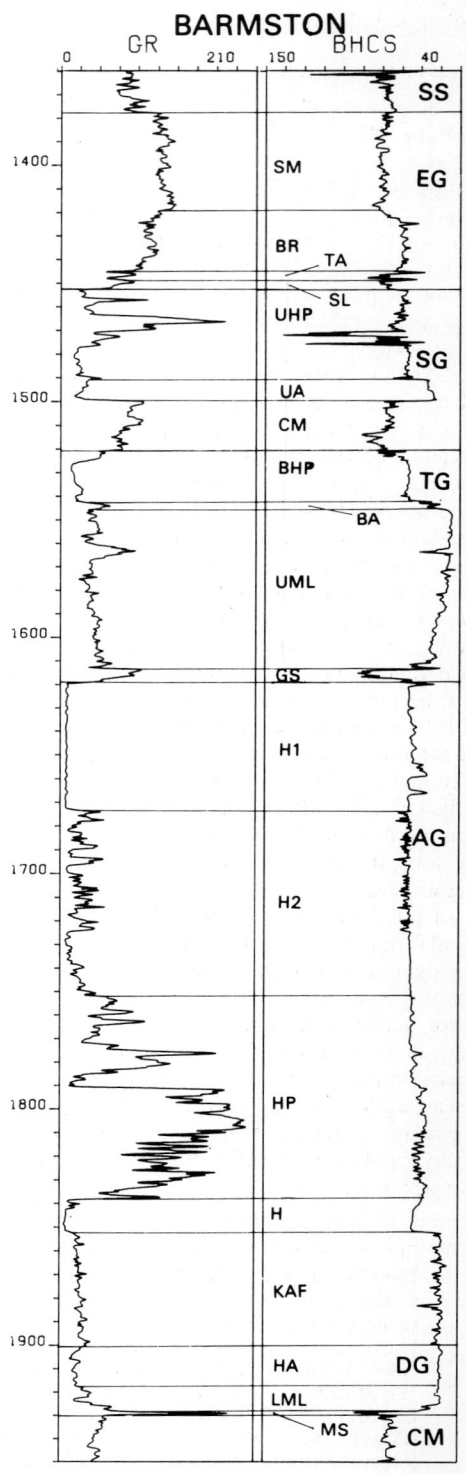

geographical area and crossed so many international boundaries.

Triassic log signatures

The Triassic succession in Britain is divided, in upward sequence, into three major lithostratigraphic units of variable thickness: the Sherwood Sandstone Group, the Mercia Mudstone Group and the Penarth Group (Warrington *et al.* 1980). The Sherwood Sandstone Group consists mainly of red, yellow and brown sandstones, with subordinate pebbly sandstones. The Mercia Mudstone Group is composed mainly of red mudstones, silty mudstones or siltstones with subordinate green beds, halite-bearing units and thin beds or nodules of sulphate evaporites: the highest unit is the Blue Anchor Formation, a series of grey, dark grey and green mudstones and siltstones. The Penarth Group comprises a lithologically variable succession of mainly shales, mudstones, siltstones and limestones. Basal Lias beds that underlie strata of the *Psiloceras planorbis* Zone (Hettangian Stage, Jurassic)—that is, the Liassic shales and limestones known locally as the *Ostrea* Beds or pre-*planorbis* Beds—are now classified with the Triassic (e.g. Warrington *et al.* 1980, Cope *et al.* 1980). However, and because macrofossils are not normally available from boreholes, the top of the Triassic is taken at the top of the Penarth Group in this account. This is the level of the top of the Langport Member (Lilstock Formation) or White Lias which is commonly a major mapping horizon on seismic sections; it is also commonly a major geophysical log marker.

In all three major lithostratigraphic divisions, local formations are named. These commonly form strong lithological contrasts with the overall, gross lithologies of the Group in which they occur, or form surface-mappable features. For the most part, however, it is difficult to surface map Triassic rocks or subdivide the Triassic sequence in detail because of poor exposure which makes it difficult to recognize detailed and subtle changes in lithology. Even cored boreholes give only limited help. Geophysical log records, however, pick out very subtle

variations in the sequence and allow detailed subdivision.

Balchin & Ridd (1970) used geophysical logs to correlate in detail the Mercia Mudstone Group sequences of eastern England and recognized regional gamma ray markers which indicated that some major lithostratigraphic units may be diachronous. Similar work using offshore data (Geiger & Hopping 1968, Brennand 1975) demonstrated that detailed Triassic correlations could be achieved across the southern North Sea linking the classic German units with those of the British mainland. Detailed variations of stratigraphy were elucidated and isopach and palaeogeographical maps produced.

Lott *et al.* (1982) described the concealed Mercia Mudstone Group of the western Wessex Basin using wireline logs. They recognized six lithostratigraphical divisions where surface mapping at the western margin of the basin distinguished only two continuous ones, a red 'marl' sequence below and relatively thin grey and green mudstones (Blue Anchor Formation) above. Although the North Curry Sandstone and its equivalents (occurring with the red 'marls') can be recognized at the surface they cannot be mapped continuously.

Southern England

For the Triassic of southern England three boreholes are illustrated here: the Nettlecombe sequence, which is typical of that in the Wessex Basin, is correlated with the BGS Burton Row (Brent Knoll) Borehole section drilled in the Central Somerset Basin and the Cooles Farm sequence in the southern part of the Severn Basin (Figs 25, 26). On the basis of the wireline logs the Sherwood Sandstone Group is divisible into three units designated SS1, SS2 and SS3 in ascending sequence. Subdivision SS1 combines low gamma ray values and relatively high velocity values (low Δt); except in the case of the Burton Row sequence the logs are not serrated but show relatively constant values betokening a rather massive sequence with poor bedding. The interval represents the basal Sherwood Sandstone Group pebble

---

FIG. 24. Permian (and Westphalian) log signatures: eastern England. CM, Coal Measures; ?THC, ?Top Hard/Barnsley Coal; ?AMB, ?Aegiranum Marine Band; SC, sandstone; LP, Lower Permian; S3, fluvial sandstone and conglomerate; S2, aeolian/fluviatile sandstone; S1, sandstone reworked by Zechstein transgression; UP, Upper Permian; DG, Don Group; MS, Marl Slate; LM, Lower Marl; LML, Lower Magnesian Limestone; HA, Hayton Anhydrite; MM, Middle Marl and anhydrite; AG, Aislaby Group; KAF, Kirkham Abbey Formation; H, Halite; HP, Halite and polyhalite; H2, Halite and minor marl; H1, Halite and minor anhydrite; TG, Teesside Group; GS, Grauer Salzton; UML, Upper Magnesian Limestone; BA, Billingham Main Anhydrite; BHP, Boulby Halite and Potash; SG, Staintondale Group; CM, Carnallitic Marl; UA, Upper Anhydrite; UHP, Upper Halite and Potash; EG, Eskdale Group; SL, Sleights Siltstone; TA, Top Anhydrite; BR, Bröckelschiefer; SM, Saliferous Marl; SS, Sherwood Sandstone Group.

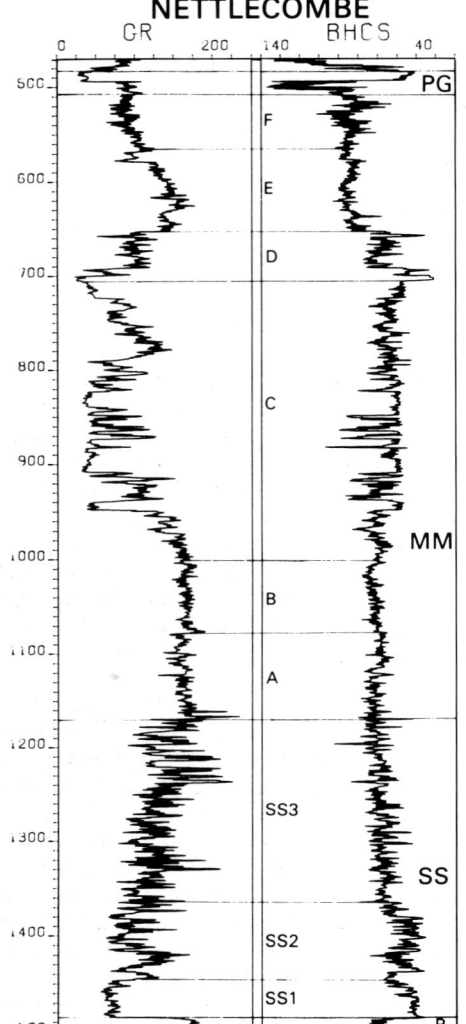

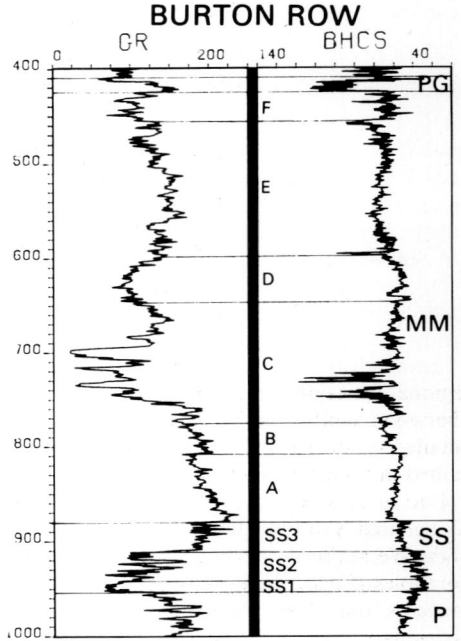

FIG. 25. Triassic log signatures. P, Permian; SS, Sherwood Sandstone Group including divisions 1 to 3; MM, Mercia Mudstone Group including divisions A to F; PG, Penarth Group.

beds and conglomerates. The Burton Row cores for this interval confirm that the sequence there comprises alternating sandstones and pebble beds, not a massive basal conglomerate as in the other illustrated sections. Subdivision SS2 is character-ized by higher gamma ray and slightly lower velocity values and a much more serrated log signature; in all cases the values (gamma ray and Δt) increase slightly up the sequence. The interval represents a predominantly sandstone sequence with some pebble and siltstone beds. Subdivision SS3 is essentially similar to SS2 in log character except that gamma ray and interval transit time (Δt) values increase slightly at the boundary and continue to increase up the sequence. The only

succession with a significant change from SS2 to SS3 is that proved at Burton Row. In this borehole the cores confirm the lithological change from predominantly siltstones above 906 m depth into sandstones below that depth. So marked is the lithological change that on this basis alone the Mercia Mudstone Group–Sherwood Sandstone Group boundary was placed here (Whittaker & Green 1983). Only with the hindsight of later regional log marker correlation is it possible to suggest that the siltstone interval recorded between the depths of 882 m and 906 m at Burton Row might be the stratigraphical equivalent of SS3 and therefore ought to be classified with the Sherwood Sandstone Group. The significance of this to

TABLE 1

Burton Row Borehole (Whittaker & Green 1983) Lithological divisions from cores		WIRELINE LOG divisions (Lott *et al.* 1982)
	Thickness (m)	
7 Blue Anchor Formation	34.52	F
6 Red mudstone (marls)	142.01	E
5 Red and purple siltstones	42.28	D
4 Upper evaporitic red siltstones	52.79 ⎱	C
3 Lower evaporitic red siltstones (including Somerset Halite Formation)	81.24 ⎰	
2 Red siltstones	106.80	A and B
1 Red siltstones and sandstones	24.45	SS3

regional facies analysis is obvious. Analysis of the Sherwood Sandstone Group data then allows more detailed subdivision of the sequence than might be gained in poorly exposed terrain.

Lott *et al.*'s (1982) six stratigraphical divisions (designated A to F) of the Mercia Mudstone Group of the Wessex Basin can be distinguished in all three boreholes (Figs 25, 26), and of particular significance is the close identification of lithological divisions in the Burton Row cores (the borehole was cored throughout) with the divisions based upon the wireline log stratigraphy (Table 1).

Units A and B are not very distinct from each other lithologically. Unit C, however, encompasses the salt-bearing parts of the Mercia Mudstone Group sequence in parts of the Wessex Basin and in the Central Somerset Basin. The halite beds in the Nettlecombe section are marked by low gamma ray and high velocity values. The equivalent salt beds are marked by similar log character in the Burton Row Borehole but the sonic log is affected by cycle-skipping and some low velocity values because of hole rugosity in the vicinity of the soluble salt beds. Although unit C of the Cooles Farm sequence is not saliferous the log character enables immediate correlation of this evaporitic silty mudstone sequence with the salt beds in more basinal sequences. Unit D shows a general decrease of gamma ray and increase of velocity values compared with the non-saliferous higher parts of unit C; at the base of unit D is a laterally persistent log marker which is commonly dolomitic and very widespread geographically. Unit E shows higher general radioactivity values and lower velocities and has a characteristic 'waisted' signature. Unit F with decreasing gamma ray but increasing velocity values and a strongly serrated character indicates the rapidly alternating siltstone-mudstone sequence of the Blue Anchor Formation (Tea Green Marl and Grey Marl) with anhydrite beds and nodules.

The Penarth Group, comprising the Westbury Formation below with the Lilstock Formation above, commonly forms a significant geophysical log marker. The black shales of the Westbury Formation are radioactive and thus show high gamma ray counts, and are characterized by low velocity values (see Fig. 27). In southern England the sequence comprises alternating black shales and limestones or calcareous sandstones; under optimum logging conditions (good hole and appropriate logging speeds), the lithological sequence can be determined very accurately on a bed-by-bed basis.

The dense, compact and porcellanous limestones of the Langport Member at the top of the Lilstock Formation contrast markedly in lithology and physical properties with the black shales of the Westbury Formation. Characteristically, the Langport Member ('White Lias') shows very low gamma ray values but very high velocities. Log signatures indicate that in the Nettlecombe and Cooles Farm boreholes it comprises a fairly massive dense limestone sequence with thin mudstone or calcareous mudstone partings like that of the Dorset coast. The cored Burton Row sequence, however, has a very thin Langport Member identical to that of the Somerset coast. All the major lithostratigraphical divisions of the Penarth Group can thus be recognized on the geophysical logs in southern England.

The Midlands and northern England

The lack of suitable released data forbids a detailed treatment of north-west England (but see Colter 1978), the Midlands and north-east England. The last two areas are marginal to the North Sea Basin and important for correlation with the offshore sequences. The area around Nottingham has been described by Elliott (1961). The released well illustrated here (Lockton 8: Fig. 26) raises several interesting points. The Sherwood Sandstone Group (Bacton Group) is readily recognized and interpreted from the geophysical logs and

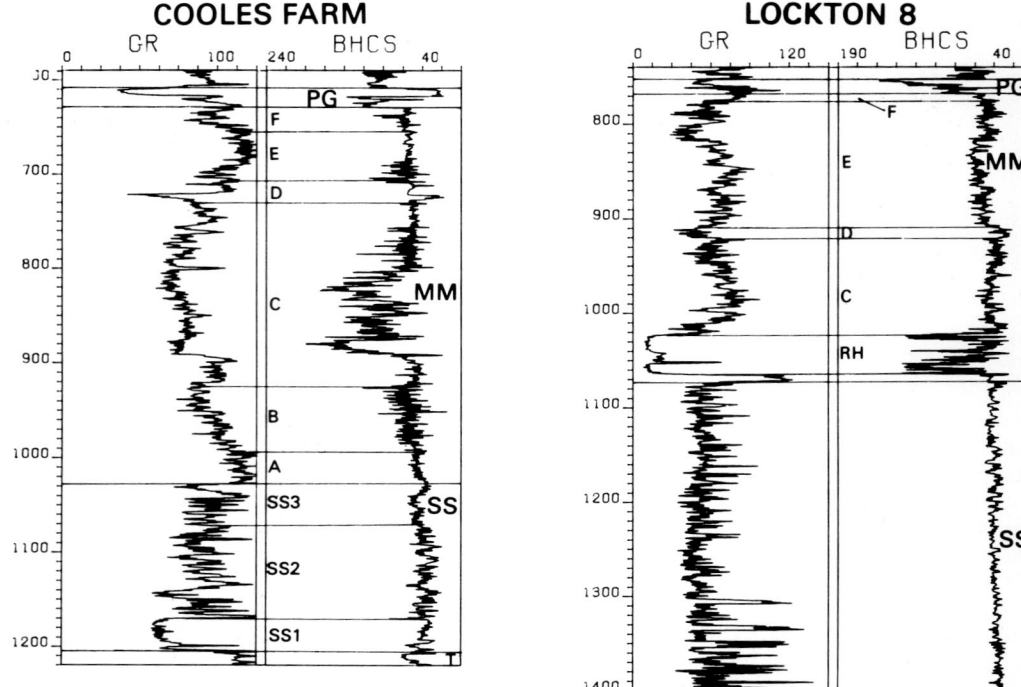

FIG. 26. Triassic log signatures. T, Tremadoc; SS, Sherwood Sandstone Group including divisions 1 to 3;
MM, Mercia Mudstone Group including divisions A to F; RH, Röt Halite; PG, Penarth Group.

correlates with North Sea sections presented by Brennand (1975); both the Bunter Shale Formation and the Bunter Sandstone Formation are identifiable. Correlation of the Mercia Mudstone Group (Haisborough Group of North Sea), however, is less straightforward in this northern area than in the southern area described above. The log signatures of Lockton 8 suggest correlation of the higher part of the Mercia Mudstone Group as shown in Fig. 26, with the basal dolomitic unit of unit D being equivalent to the basal unit of the offshore Keuper Anhydritic Member (of the Triton Anhydritic Formation). However, on the basis of local correlation of landward and offshore data, the halite beds at the base of the Lockton 8 Mercia Mudstone Group sequence do not equate with the Carnian age salts of unit C or the offshore Keuper Halite Member, but with the older Röt Halite Member of Scythian age (Warrington et al. 1980). There is danger therefore in a simple long-range correlation of stratigraphical information derived from geophysical logs just as there is with such information derived solely from lithological records. An important point arising from this conclusion, though, is the probability that much of the middle section of the Mercia Mudstone Group

sequence is missing or extremely condensed in Lockton 8.

Lower Jurassic log signatures

The English Jurassic lithostratigraphy and biostratigraphy are known extremely well as a result of nearly two hundred years of research on inland and coastal exposures. There are rapid vertical changes in rock types, and the sequence is particularly amenable to study by means of downhole geophysical logs controlled by well-understood surface geology, palaeontological research and cored borehole sections.

The Lias is subdivided into the Lower Lias, the Middle Lias and the Upper Lias: the stratigraphy has been reviewed by Cope et al. (1980). Coastal sections provide the best exposures, and numerous lithostratigraphical divisions are recognized on the Dorset and Yorkshire coasts. On the Somerset and Glamorgan coasts significant (usually higher) parts of the succession are missing.

Few papers on onshore Lower Jurassic geology incorporate stratigraphically useful wireline log information, but Horton & Poole (1977) have described three widespread and persistent electrical

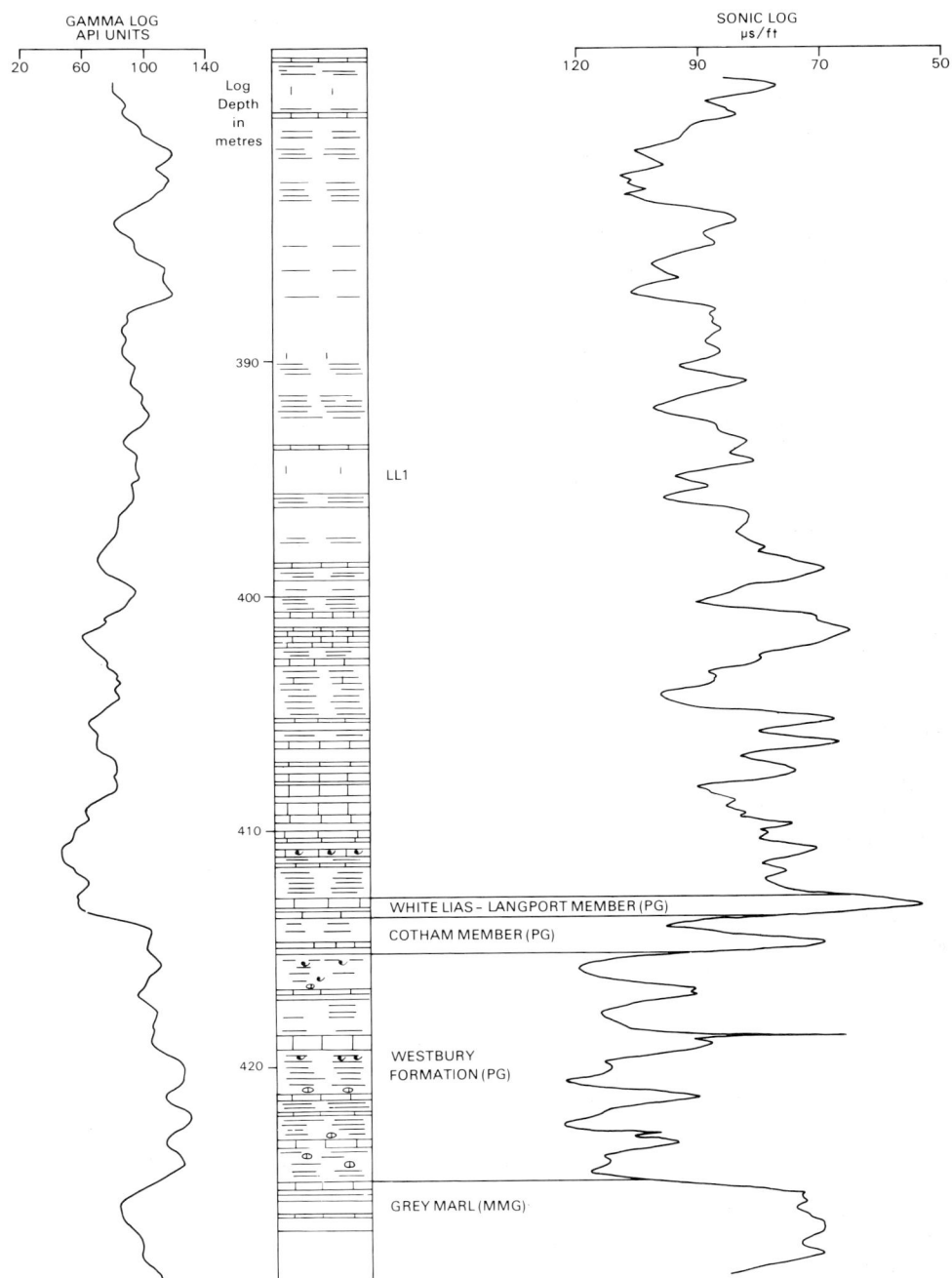

FIG. 27. Detailed correlation of Liassic and Penarth Group lithologies with log signatures in the Burton Row Borehole. MMG, Mercia Mudstone Group; PG, Penarth Group; LL1, lowest division of Lower Lias.

resistivity markers in the Lower Lias of Oxfordshire. Recognition of two of these markers was
extended to the BGS Barby Borehole sequence of
Northamptonshire by Ambrose & Ivimey-Cook
(1982).

Lower Lias

From Dorset to south Humberside the Lower
Lias is predominantly argillaceous but with a
significant calcareous content particularly in its
lower part. It is best known on the Dorset coast
from the detailed research of Lang and his co-
authors. The five lithological divisions recognized
there (see below) can only be mapped at surface for
a very short distance inland. Generally it is only
possible to map two lithological divisions, a lower

limestone-rich unit (commonly designated Blue
Lias) and an upper clay-rich unit (usually designated Lower Lias Clays). In the subsurface of
southern England, however, geophysical logs
enable precise delimitation of the five lithological
divisions of the Dorset coast (Figs 28 to 32).

To demonstrate the validity of the larger scale
correlations the cored BGS Burton Row Borehole,
Somerset (Whittaker & Green 1983) is referred to
here. In Fig. 27 the detailed lithological record of
part of the Lower Lias sequence from core descrip-
tion is compared with the gamma ray and sonic
logs run over the same interval. A precise correlation is apparent between the alternating mudstones
and limestones of the Blue Lias sequence and the
log signature. It is striking that beds of only 20 cm

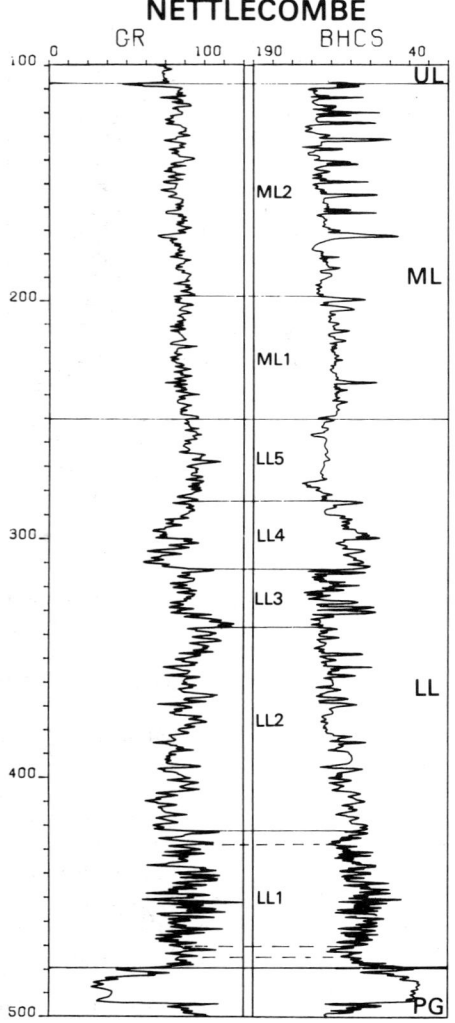

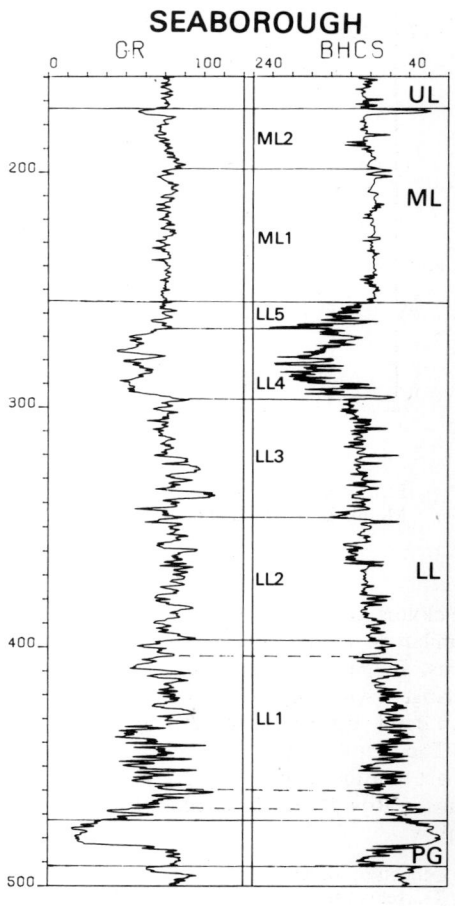

FIG. 28. Lower Jurassic log signatures. PG,
Penarth Group; LL, Lower Lias including
divisions 1 to 5; ML, Middle Lias including
divisions 1 and 2; UL, Upper Lias.

WINTERBORNE KINGSTON

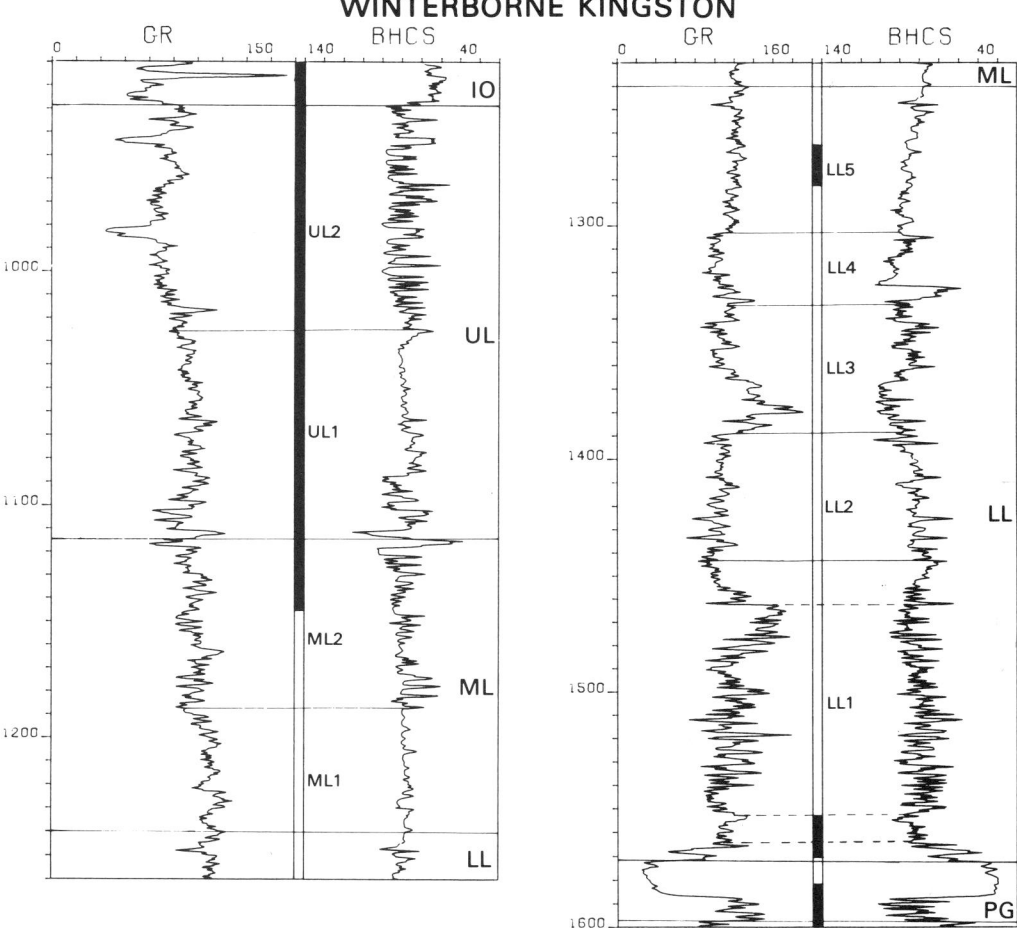

FIG. 29. Lower Jurassic log signatures. PG, Penarth Group; LL, Lower Lias including divisions 1 to 5; ML, Middle Lias including divisions 1 and 2; UL, Upper Lias including divisions 1 and 2; IO, Inferior Oolite.

thickness can be resolved by the wireline logs; similar detail is available throughout the Lower Lias. In parts of the sequence there are subtle changes in lithology recorded by the wireline logs but barely discernible to the eye.

The lithostratigraphical divisions of the Dorset coast and the wireline log units recognized in the Burton Row Borehole sequence (Fig. 30) are given below.

Lithostratigraphical units (Dorset coast)	Geophysical log units (Burton Row Borehole)
Green Ammonite Beds	LL 5
Belemnite Marl	LL 4
Black Ven Marl	LL 3
Shales-with-beef	LL 2
Blue Lias	LL 1

The log units are constrained by the ammonite stratigraphy of the Burton Row section (for details of the lower part of the sequence see Ivimey-Cook & Donovan 1983) which correlates well with the Dorset coast.

Lower Lias unit LL 1 is characterized by an extremely serrated and spikey signature reflecting the rapid limestone–mudstone alternations of the Blue Lias. However, an 'average' line (lower frequency component of the signal) through the spikes has a sinusoidal shape and indicates predominance of mudstone over limestone or vice versa. By this means it is possible to further subdivide unit LL 1. One demonstrable example of the validity of this is the recognition of a predominantly mudstone subdivision near the base of the

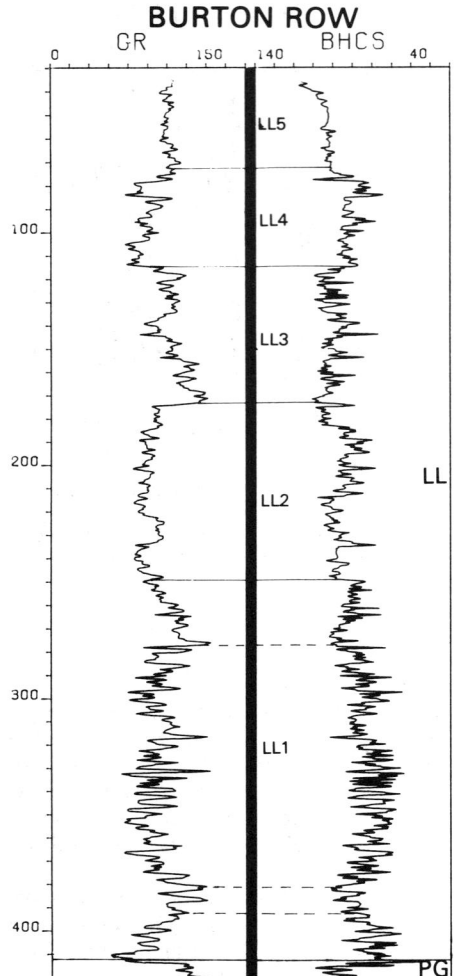

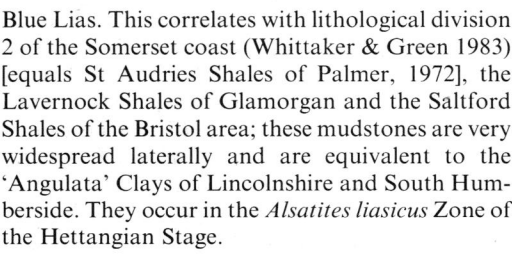

FIG. 30. Lower Jurassic log signatures. PG,
Penarth Group; LL, Lower Lias including
divisions 1 to 5; ML, Middle Lias including
divisions 1 and 2; UL, Upper Lias including
divisions 1 and 2; IO, Inferior Oolite.

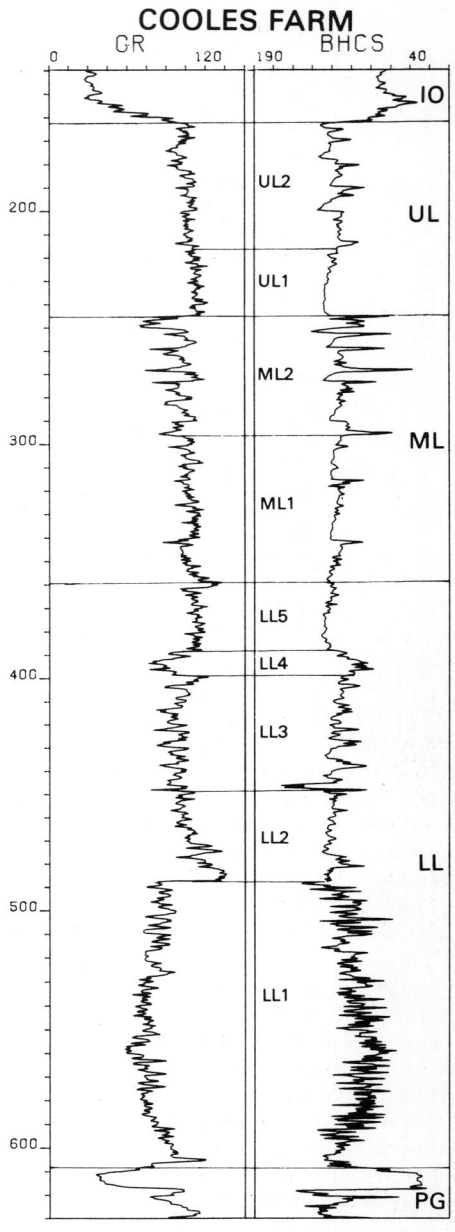

Blue Lias. This correlates with lithological division
2 of the Somerset coast (Whittaker & Green 1983)
[equals St Audries Shales of Palmer, 1972], the
Lavernock Shales of Glamorgan and the Saltford
Shales of the Bristol area; these mudstones are very
widespread laterally and are equivalent to the
'Angulata' Clays of Lincolnshire and South Humberside. They occur in the *Alsatites liasicus* Zone of
the Hettangian Stage.

Unit LL 2 has a less serrated log signature than
unit LL 1 and is marked by a slightly lower gamma
ray value and lower velocity value. In carefully
logged sequences there is a sharp decrease in
velocity at this level.

At its base, unit LL 3 has higher gamma ray and
lower velocity values which decrease and increase,
respectively, up the LL 3 sequence.

Unit LL 4 is an important marker. Gamma ray

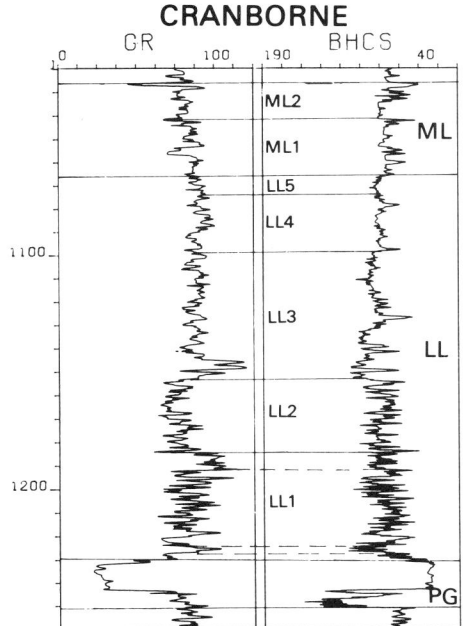

FIG. 31. Lower Jurassic log signatures. PG, Penarth Group; LL, Lower Lias including divisions 1 to 5; ML, Middle Lias including divisions 1 and 2; UL, Upper Lias including divisions 1 and 2; MJ, Middle Jurassic.

values decrease while velocity values increase. The interval is widely recognized, and on the basis of the geophysical logs is divisible into three subunits in the south. Three highly calcareous beds are marked by sonic log peaks; the lower two of these appear to correlate with the '85' and '70' Marker Members recognized by Horton & Poole (1977) in Oxfordshire. Similar correlation is possible with beds near the level of the Pecten Ironstone in Lincolnshire and Humberside (Fig. 31).

On the basis of log character, unit LL 5 shows a poorly serrated signature with fairly constant gamma ray values (invariably higher than unit LL 4) but rather more variable velocity values.

Correlation of the Yorkshire Basin sequence with that of southern England is possible, despite the siltier lithological development, using results from the cored and geophysically logged Felixkirk Borehole (Powell 1984). All of the Lower Lias units are present (Fig. 32) and the log correlation is confirmed by the ammonite stratigraphy. Noteworthy is the precise correlation of the Redcar Mudstone Formation with the Lower Lias, and of individual lithostratigraphical units of the Yorkshire Basin with those of Dorset.

The examples cited here demonstrate the enormous potential of geophysical logs in stratigraphical studies of what are commonly thought to be monotonous, uniform sequences. It is clear that Lower Lias sequences are capable of much finer and more detailed subdivision that has hitherto been attempted. Long natural exposures or borehole cores with good ammonite control can be integrated with the geophysical log stratigraphy to give excellent biostratigraphical control in properly logged, uncored borehole sections in the subsurface.

It is becoming possible to recognize meso- and mega-cycles of sedimentation as well as the obvious microcycles at the scale of the Blue Lias rhythms. Close correlation using wireline logs will enable a much tighter integration of palaeontological and lithological information and may lead to the precise positioning of ammonite zones in poorly fossiliferous sequences. Such work would have an important bearing on the distribution of ammonite taxa in time and space.

Middle Lias

The Middle Lias shows a broad two-fold division; a lower unit of mudrocks passing up into siltstones or sandstones is overlain by the prominent Marlstone Rock Bed. The latter is a hard, commonly sandy limestone which in the Midlands becomes strongly ferruginous in places before passing to ironstones and shales (Cleveland Ironstone Formation) in Yorkshire. The two divisions can be surface-mapped over much of the outcrop and comparable units (ML 1 and ML 2) are commonly recognizable on geophysical logs (Figs

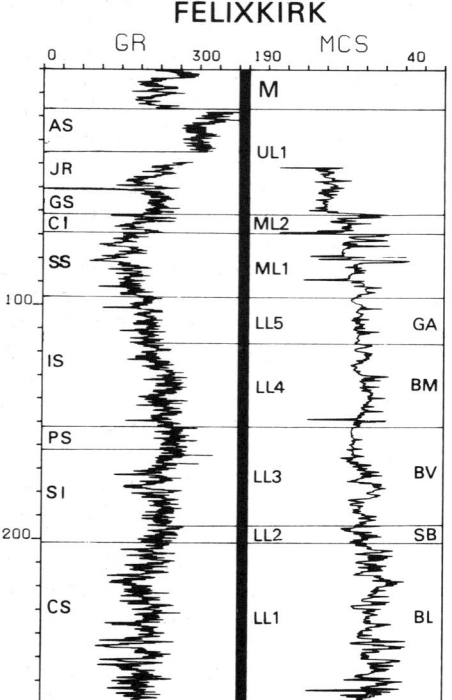

FIG. 32. Lower Jurassic log signatures and relationship between lithological divisions recognized in Yorkshire (left) and south-west England (right). LL 1 to 5, divisions of the Lower Lias; BL, Blue Lias; SB, Shales with Beef; BV, Black Ven Marl; BM, Belemnite Marl; GA, Green Ammonite Beds; CS, Calcareous Shales; SI, Siliceous Shales; PS, Pyritous Shales; IS, Ironstone Shales; ML 1 and 2, divisions of the Middle Lias; SS, Staithes Sandstone Formation; CI, Cleveland Ironstone Formation; UL1, Lower division of the Upper Lias; GS, Grey Shale Member; JR, Jet Rock Member; AS, Alum Shale Member; M, Middle Jurassic; MCS, Multi-Channel Sonic Log.

England is taken at the change between relatively uniform signature below and a much more spikey signature above. The geophysical log sequence culminates at the top of the Middle Lias in sharp peaks of low gamma ray and high sonic velocity values. This is taken as the level of the Marlstone Rock Bed or the Junction Bed.

Upper Lias

The Upper Lias sequence is well known from surface geological studies and from cored, but rarely logged, boreholes (Figs 29, 30, 31). The sequence varies in lithology and thickness, with diachronism of lithological units and non-sequences being particularly characteristic. However, it commonly comprises a lower argillaceous lithological unit (of variable age) overlain by an arenaceous unit (also of variable age) sometimes capped by calcareous beds. Such sequences occur in southern England, the Cotswolds and Yorkshire.

Geophysically logged Upper Lias borehole sequences are commonly not constrained by extensive coring runs yielding diagnostic fossils (though many cores have been obtained from the Upper Lias sands of the Wytch Farm area). Despite this the log records show that the Upper Lias is divisible into two units (UL 1 and UL 2). The lower unit (UL 1) has a less serrated log signature with poorer bed definition than the upper unit (UL 2) and overall slightly higher gamma ray but lower velocity values. Unit UL 2 in places has a very spikey and 'noisy' signature depending upon the presence of hard, calcareous doggers in the southern England sequences. Throughout UL 2, however, gamma ray and sonic velocity values are fairly uniform. The two units correspond with the argillaceous and arenaceous units at outcrop.

Middle Jurassic log signatures

The English Middle Jurassic rocks show a passage from terrigenous, even paludal and fluvial, sediments in the north through the brackish and lagoonal succession of the Midlands to the inner-shelf mudstones and carbonate sequences of the south. The facies thus contrast markedly with the dominantly mudstone sequences of the underlying Lias and overlying Upper Jurassic successions. In places a cyclic repetition of facies occurs which, when combined with lateral changes in facies, thickness and widespread non-sequence, renders stratigraphical correlation difficult. Generally speaking, the most dramatic thickness changes occur across the location of presumed growth faults and result in more complete sequences

28 to 32). The classic Middle Lias sequence of the Dorset coast can be further subdivided (Day 1863, Howarth 1957).

The geophysical log signatures of borehole sequences inland show features that might be anticipated by comparison with the coastal outcrop. The logs show relatively constant gamma ray and sonic values in general with slight decreasing gamma and increasing sonic values up the sequence. However, sharp spikes occur, representing hard sandy and calcareous bands in the lower part and similar bands or doggers in the upper part. The boundary between the two Middle Lias geophysical log units (ML 1 and ML 2) in the south of occurring on the downthrow side and attenuated sequences on the upthrow side of such faults.

The carbonate succession of southern England is divided into a lower Inferior Oolite Group, mostly of Bajocian (including Aalenian) age and an upper, Great Oolite Group all of which is of Bathonian age with the exception of the highest member, the Upper Cornbrash. In Yorkshire the Ravenscar Group, which is now known to be entirely of Bajocian age, rests on the Lower Bajocian (Aalenian) Dogger Formation and again is overlain by the Callovian, Upper Cornbrash.

Inferior Oolite and Ravenscar Groups (including Dogger)

The geophysical log signature of the Inferior Oolite usually allows its easy separation from the underlying Lias and overlying Fuller's Earth. At the lower junction there is invariably a sharp rise in sonic velocity and corresponding decrease in gamma ray values, a contrast that is particularly marked where the underlying Lias is developed in mudstone facies (Figs 33, 34). Even where the Lias is in its Upper Liassic sand facies the contrast is easily detectable for the latter tends to yield a spikey sonic log signature where less well cemented sandstones alternate with hard, well cemented doggers. A marked decrease in sonic velocity and corresponding increase in gamma ray values invariably marks the top of the Inferior Oolite. The persistence of high gamma ray values and low sonic velocity in the overlying Fuller's Earth contrasts strongly with the Inferior Oolite log motif and allows easy separation and demarcation of the rock bodies (Figs 33, 34). Something of the difficulty in internally subdividing and correlating the Inferior Oolite is apparent from considering the sequence from two fully cored boreholes in the Wessex Basin (Fig. 33). In both, the Inferior Oolite as a whole forms a distinctive segment of the log motif. Both show similar, average sonic velocity, though the gamma ray values at Winterborne Kingston are higher than those obtained at Seabarn Farm. The extremely fine serration of the bulk of the succession at Seabarn Farm reflects the rapid alternations of more or less calcareous siltstones whereas the coarser log peaks at Winterborne Kingston reflect the thicker limestone posts encountered there. Prominent, high gamma ray values occur at Winterborne Kingston and reflect the presence of substantial interbeds of calcareous mudstone and siltstone. The most prominent gamma ray peak, however, is a response from the level with abundant phosphatized pebbles around 916 m depth. Such a peak also occurs at Seabarn Farm (Fig. 33) and is

widespread in the Wessex Basin where it is also associated with major non-sequence (Penn 1982).

On the northern flank of the Wessex Basin, the Inferior Oolite sequence is of comparable thickness to that at Winterborne Kingston but over a substantial area to the north of the Mendip 'High' only Upper Inferior Oolite is present (Fig. 34). The sequence is well known and has been traced over a large area at outcrop. The geophysical log signature is correspondingly easily identified though difficulties may arise when Lower and Middle Inferior Oolite strata are present between the Lias and the Upper Inferior Oolite. Thus the Upper Trigonia Grit may usually be recognized by its combination of a minor gamma ray peak and prominent high velocity peak in the basal few metres of the sequence. The high gamma ray values are thought to reflect the prominent marly content of the coarsely bioclastic limestones which are frequently recrystallized during hard-ground formation.

The succession at Cooles Farm differs from that at Devizes (Fig. 34) in that Lower Inferior Oolite appears between the unconformity at the base of the Upper Inferior Oolite and the Lias. Here the gamma ray values decrease less rapidly upwards from those of the underlying Lias and the sonic velocity increases correspondingly upwards. The gamma ray signature here resembles that logged at Stowell Park (Green & Melville 1956). The low sonic velocities are thought to reflect the friable nature of the sandy, Scissum Beds.

North of the London Platform area the principal features of the geophysical log signature of the typical East Midlands Bajocian succession of Northampton Sands overlain by Grantham Formation and Lincolnshire Limestone is shown by the fully cored Nettleton Bottom Borehole (Fig. 34). The borehole is of particular interest in that it shows a southward extension of the Yorkshire Ravenscar Group facies intercalated between the top of the Lincolnshire Limestone and the base of the overlying Great Oolite Group. The gamma ray curve of the Northampton Sands is one of moderate value; minor peaks and a more prominent peak occur just above the base. The sonic log shows moderately high sonic velocity with minor and major low values corresponding to the peaks of the gamma ray motifs. Thin, higher velocity peaks also occur. These features reflect the fact that the sandstones are interbedded with minor and major levels of mudstones and siltstones which are commonly carbonaceous as well as being sporadically tightly cemented with calcite or siderite. The overlying Grantham Formation is dissimilar only in that the gamma ray peaks and low velocity sonic

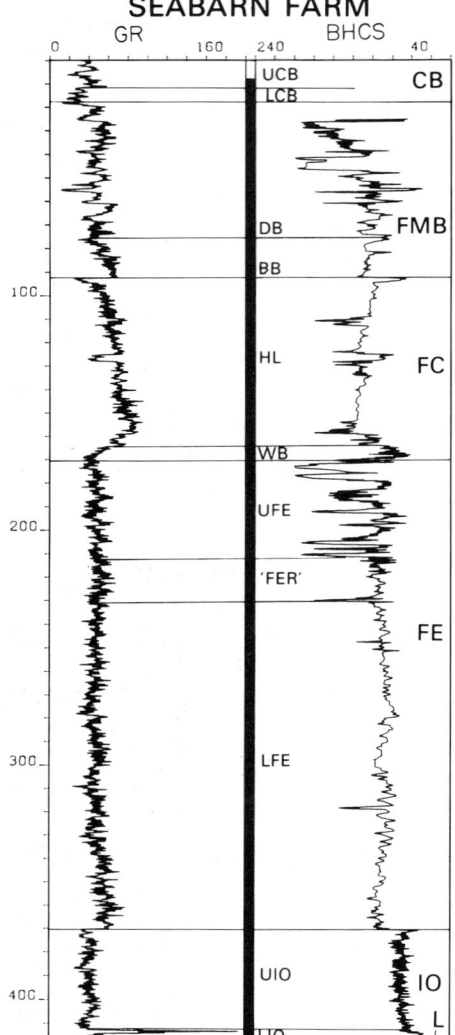

FIG. 33. Middle Jurassic log signatures: Wessex Basin. L, Lias; IO, Inferior Oolite; LIO, Lower Inferior Oolite; MIO, Middle Inferior Oolite; UIO, Upper Inferior Oolite; FE, Fuller's Earth; LFE, Lower Fuller's Earth including beds 1 to 9 of Penn *et al.* (1979); FER, Fuller's Earth Rock; 'FER', Fuller's Earth Rock equivalent; UFE, Upper Fuller's Earth; FC, Frome Clay; WB, Wattonensis Beds; HL, Hebridica Lumachelle; FMB, Forest Marble; BB, Boueti Bed; DB, Digona Bed; CB, Cornbrash; LCB, Lower Cornbrash; UCB, Upper Cornbrash; KB, Kellaways Beds. Gamma ray log in counts/s.

values are higher and lower respectively. In addition, the carbonaceous mudstones which give rise to them ocupy a larger part of the sequence than do the intervening siltstones. This Lower Bajocian sequence is quite characteristic of the East Midlands.

The Lincolnshire Limestone signature falls into three parts. In the lower part, the gamma ray values decrease upwards but show angular peaks alternating with pronounced low velocity values (Fig. 34). This reflects the fact that the lower part of the limestone contains interbedded, hard, sometimes

porcellaneous limestones and softer silty mudstones. In the middle part the gamma ray values continue to decrease upwards and maintain the moderate strength, angular peaks, but there is commonly a very pronounced gamma ray peak. The sonic log, however, shows a 'ratty' signature of consistently high sonic velocity. It is thought that these features reflect the alternating cementstones and hard, interbedded shales commonly occurring in the median part of the Lincolnshire Limestone and called Kirton Cementstones. The large gamma ray peak may then correspond to the level of the

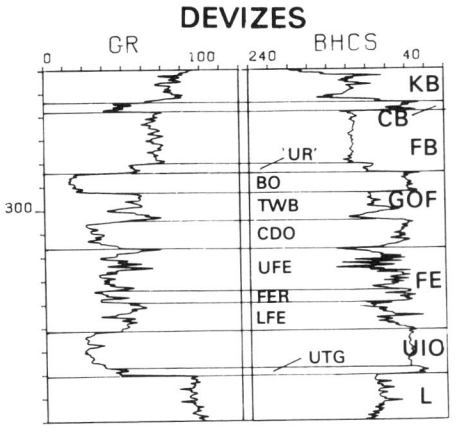

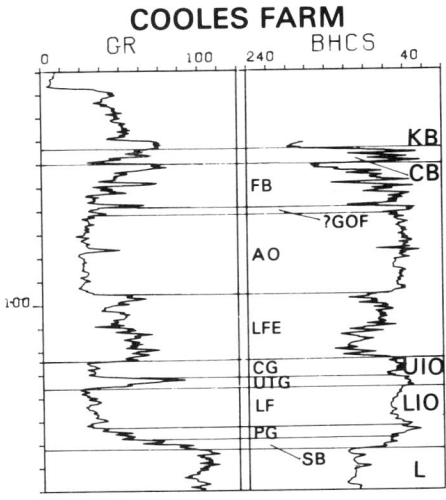

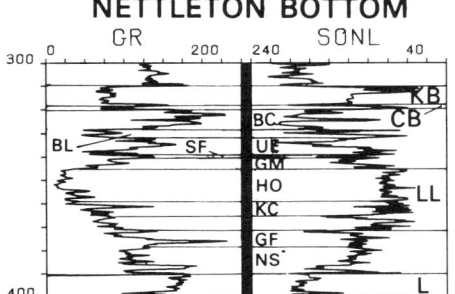

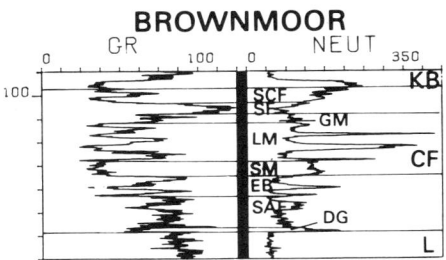

FIG. 34. Middle Jurassic log signatures. L, Lias; LIO, Lower Inferior Oolite; SB, Scissum Beds; PG, Pea Grit; LF, Lower Freestone; NS, Northampton Sands; GF, Grantham Formation; DG, Dogger; SAF, Saltwick Formation; EB, Ellerbeck Formation; LL, Lincolnshire Limestone; KC, Kirton Cementstones; HO, Hibaldstow Oolite; CF, Cloughton Formation; SM, Sycarham Member; LM, Lebberston Member; GM, Gristhorpe Member; SF, Scarborough Formation; UIO, Upper Inferior Oolite; UTG, Upper Trigonia Grit; CG, Clypeus Grit; SCF, Scalby Formation; FE, Fuller's Earth; LFE, Lower Fuller's Earth; FER, Fuller's Earth Rock; UFE, Upper Fuller's Earth; AO, Athelstan Oolite; UE, Upper 'Estuarine' Beds; BL, Blisworth Limestone; BC, Blisworth Clay; GOF, Great Oolite Formation; CDO, Combe Down Oolite; TWB, Twinhoe Beds; BO, Bath Oolite; FB, Forest Marble; 'UR', Upper Rags Equivalent; CB, Cornbrash; KB, Kellaways Beds. Both logs at Brown Moor and gamma ray log at Nettleton Bottom in counts/s.

Kirton Shale although the presence of the latter was not recognized by Bradshaw & Penney (1982, p.121). The uppermost part of the signature usually has the lowest gamma ray values and often shows a slight increase to the top of the formation. It corresponds to the clean, well sorted, oolitic limestones of the upper part of the Lincolnshire Limestone, the Hibaldstow Oolite. In these, marly partings occur and the lower part tends to be somewhat argillaceous.

All three subdivisions of the Lincolnshire Limestone log signature can usually be recognized in upward sequence but lateral facies change as well as

internal attenuation and overstep means that their boundaries may be diachronous. Nevertheless in many places it is possible to recognize the subdivisions of Ashton (1980).

The top of the formation is usually marked by a sharp fall in gamma ray values. Above (Fig. 34) is an upwards decreasing, 'stepped' gamma ray profile and corresponding decreasing sonic log signature; the gamma ray profile marks calcareous sandstone and siltstone which became increasingly argillaceous and even coaly upwards but are interbedded with thin limestone which gives rise to the 'step' part of the profile. These beds resemble, and yield a fauna indicating thay they are the correlatives of, the Scarborough Formation.

North of the Market Weighton area, the geophysical log signature of equivalent beds is not accurately known. The cored borehole at Brown Moor (Fig. 34) just to the north of Market Weighton, but not within the Cleveland Basin proper, is of importance in that elements of the East Midlands facies are still present within the dominantly terrigenous sequence (Gaunt *et al.* 1980). To the base, the geophysical log signature shows a moderate neutron peak which forms a definite spike and moderately low gamma ray values. This corresponds to the sporadically calcareous sandstones of the Dogger Formation at the base of which is a sandy limestone. The overlying beds are characterized by finely serrated log signatures recording moderate to high gamma ray values and moderate to low neutron values. Near the top is a serrated minor neutron peak and corresponding low gamma ray values. This, the Saltwick Formation, comprises non-calcareous, silty mudstones interbedded with micaceous siltstones and laminated sandstones with locally abundant plant debris. The neutron peak towards the top reflects a thicker development of sandstone. Although the Brown Moor sequence is sandier, it is possible to see in these Lower Bajocian (Aalenian) sediments a similar sequence to that of their correlatives to the south of the Humber.

Above these beds the geophysical log signature is dominated by two prominent low gamma ray spikes each with corresponding moderate neutron values in a signature otherwise characterized by high gamma ray and low neutron values. This signature is that of the Ellerbeck Formation which comprises calcareous mudstones and siltstones within which are two levels of calcareous siltstone and sandstone. The lower of these is sideritic and the upper includes a thin, porcellaneous limestone. These beds can be correlated in detail across the Market Weighton area (Gaunt *et al.* 1980, p.29) to the south of which they constitute the lower, argillaceous part of the Lincolnshire Limestone (see Fig. 34). Above this line the gamma ray values fall sharply to low values characteristic of the fine-grained, micaceous and carbonaceous sandstones of the Sycarham Member. Moderate neutron values indicate some uniform porosity in these beds which have no obvious correlative in the East Midlands. The geophysical log signature of the sandstone body is capped by a very thin, needle-like spike of low gamma ray values which is mirrored by the neutron curve. This signature corresponds to the basal, often loosely compacted (and therefore porous), oolitic limestones and immediately over-

lying mudstone of the Lebberston Member. Above here the gamma ray curve displays a repetitive rapid decrease in values, followed by more gradually increasing values. There is a correspondingly rapid increase in neutron values followed by a more gradual decrease. This signature corresponds to the repetitive sequence of the remainder of the Lebberston Member in which mudstones pass up through sporadically calcareous sandstones to thin, loosely compacted, oolitic limestones. The limestones are overlain by silty limestones which pass upwards into mudstones. These interdigitated limestones have been correlated with higher parts of the Lincolnshire Limestone to the south (Gaunt *et al.* 1980).

The succeeding Gristhorpe Member and Scarborough Formation give rise to a log motif similar to their correlatives farther south (Fig. 34) but, to the north (Fig. 34), are overlain by Scalby Beds which give rise to a signature characterized by low gamma ray and upward increasing neutron values reflecting the presence of micaceous sandstones which are laminated, silty and less porous to the base. Within this sandy sequence, silty, carbonaceous levels give rise to higher gamma ray response.

Great Oolite Group

The Great Oolite Group comprises a complex sequence of predominantly calcareous mudstones, argillaceous, oolitic and micritic limestones and more or less calcareous sandstones. It is entirely Bathonian in age, except for the Upper Cornbrash at the top, which is Callovian. The Group has several distinctive features chief of which is, perhaps, the way in which calcareous mudstones with thin argillaceous limestones predominate in the south-west of England but are replaced north-westwards by oolitic limestones. These in turn pass into micritic limestones farther in that direction till, flanking the area of the London Platform, mudstones, fine-grained sandstones and limestones occur. North of the platform appear sandstones and mudstones thought to have been deposited in brackish to freshwater environments. A second distinctive aspect of the group is the way in which over most of England the facies boundaries migrated farther to the south and south-west through time till some of the highest beds are to be found over the whole of southern England in facies (within the Forest Marble) similar to those low in the sequence confined to the area of the London Platform. The result is that similar stratigraphical sequences occur which are of different ages from place to place. The pattern is further complicated, however, by two other features. Firstly, Bathonian sequences are invariably lithologically repetitive.

Secondly, they show considerable variation in thickness, generally associated (as in the Inferior Oolite) with growth faulting so that attenuated sequences, possibly even non-sequences, may occur on the upthrown side of such faults and fuller sequences may be found on the downthrown side. Thus it will be seen that the combination of lateral facies change and the presence of non-sequence in vertically repetitive successions may easily lead to stratigraphical miscorrelation.

The gamma ray signature of the Fuller's Earth registers higher values than those of the underlying Inferior Oolite from which it is obviously differentiated, but the values are invariably lower than those of other major Jurassic mudstones (e.g. the Lias, the Oxford Clay and the Kimmeridge Clay). In particular the gamma ray values are always considerably lower than those of the immediately overlying Frome Clay, a feature of considerable prominence in those areas in which no Fuller's Earth Rock is present (e.g. Fig. 33). This is thought to reflect the generally more uniformly calcareous nature of the Fuller's Earth by comparison with these other Jurassic mudstones. The uniformity of the mudstones is brought out by the subdued nature of the serrations of the curve in which minor peaks correspond to less calcareous beds. The sonic log shows a corresponding mildly serrated nature, but in those sequences where the Fuller's Earth Rock is absent, it is marked by the repeated occurrence of very low velocity levels in its upper part. At Seabarn Farm (Fig. 33) these commence at a level (about 211 m) some 6 m above a horizon containing abundant ornithellid brachiopods which occur in a sequence that can be regarded, therefore, as locally equivalent to the Fuller's Earth Rock. Although the succession here has not yet been worked out in detail, preliminary examination suggests that strata equivalent to the Fuller's Earth Rock probably lie between 211 m and 230 m. Thus despite some lateral facies change the three-fold division of the Upper Fuller's Earth may be traced across southern England from the clay basinal areas of the south coast to the more limestone-dominated shelf area immediately to the south of the area of the London Platform.

Lateral facies variation is most clearly brought out by the changes in sonic log signature as the Lower Fuller's Earth is traced from south to north. Thus, in the extreme south the high velocity sonic peaks are more numerous but more subdued than those occuring in thinner successions farther north (Fig. 33). In these thinner sequences, discrete, argillaceous and silty limestones give rise to the more prominent high velocity sonic spikes which contrast markedly with the lower velocities of the

intervening mudstones. Still farther north the prominent sonic spikes correspond to more calcareous levels identified (Penn *et al.* 1979) in a thin but regular sequence in which named shell-beds can be traced over large areas, e.g. between Bath and Cirencester. Study of cored boreholes, however, shows higher beds to the north to become more calcareous and include sandy limestone and sporadically argillaceous, oolitic limestones equivalent to Stonesfield Slate and Taynton Stone respectively (Fig. 34). These do not occur to the south and are assumed to be cut out by the non-sequence indicated by the widespread erosion surface at the base of the Fuller's Earth Rock. Although such stratigraphical details of the Lower Fuller's Earth have been evaluated from the study of borehole cores, it is apparent that in the fuller sequence to the south, there is a larger scale similarity in the sonic log motif which may be used in correlation (Fig. 33). For example, the lower sonic velocity of the basal 30 m at Seabarn Farm may correspond to Unit 1 and the lower part of Unit 2 as at Winterborne Kingston (Penn 1982). Similarly the overlying higher sonic velocities between that level and 305 m succeeded by a lower velocity to *c.* 280 m at Seabarn Farm may correspond to Units 3 and 4 at Winterborne Kingston. Unit 5 in both areas may be similarly identified but above this level the correlation is not clear.

The geophysical log signature of the Fuller's Earth Rock forms a prominent high velocity sonic peak and gives rise to corresponding low gamma ray values (Figs 33, 34). Usually there is a rapid change in values associated with its upper and lower boundaries reflecting both the lithological contrast with the underlying mudstones, on which the limestones rest with local non-sequence, as well as the rapid upward change to the mudstones of the Upper Fuller's Earth. In detail, the high velocity peak is found to be serrated. This, and the corresponding minor gamma ray peaks, are the log responses of the interbedded, usually shelly marls, calcareous siltstones and mudstones.

The Upper Fuller's Earth is characterized by a mildly serrated gamma ray signature in which thin, minor peaks alternate with thicker levels of lower values. These reflect the alternation of generally thinner, less calcareous beds with thicker, more calcareous beds, some of which may be argillaceous limestones (Figs 33, 34). The sonic log changes correspondingly in value but its main feature is the presence of some very low velocity levels generally corresponding to the higher values of the gamma ray curve. These distinctive responses reflect the occurrence of dark grey, sometimes black, shaley mudstones, some of which on drilling proved to be waterlogged and undercompacted. It may be that the two major calcareous levels depicted at Devizes (Fig. 34) correspond to those parts of the sedimentary cycles described in upward succession as Units 17–19 and 21 in the type area. If so then the Great Oolite at Devizes may rest unconformably on the black shales of Unit 22 as it is known to do south of Bath (Penn *et al.* 1979). To the north of the type area, the Upper Fuller's Earth passes into an oolitic limestone sequence which is again characterized by low gamma ray values and high sonic velocity. This appears to be in continuity with those limestones thought to be equivalent to the underlying Fuller's Earth Rock (Fig. 34).

The argillaceous limestones and interbedded calcareous siltstones of the basal part of the Frome Clay yield a log signature which essentially repeats that of the Fuller's Earth Rock (Fig. 33). This basal unit passes up more or less rapidly into black, shaley mudstones which give rise to a well-defined unit of low gamma ray values (some of which are the lowest encountered in the Fuller's Earth and Frome Clay) and equally pronounced low sonic velocity akin to the values encountered in the beds of similar lithology in the Upper Fuller's Earth. Above this distinctive level, the moderately serrated gamma ray curve shows steady upward decrease to the top of the formation and the moderately serrated sonic log shows similar upwards increase in sonic velocity. This steady upward change reflects a gradual upwards increase in carbonate mud and silt. Regionally, the signature depicted in Fig. 33 is very typical, widespread and easy to recognize. Within the type area of the Bathonian, the Frome Clay passes into the Great Oolite and shows the low gamma ray, high sonic velocity signatures of the Combe Down Oolite and Bath Oolite separated by the high gamma ray values and lower sonic velocity of intervening mudstones taken to be mudstones equivalent to the Twinhoe Beds (Fig. 34). To the north, towards the area of the London Platform, the entire sequence is in an oolitic limestone facies and gives rise to the appropriate signature throughout (Fig. 34). Around Cooles Farm, however, shallow drilling suggests that almost all this sequence is the lateral extension of the Combe Down Oolite and that the upper two members are absent; perhaps they are represented in the hard limestones (thought to represent a hardground complex) at the top of the sequence which gives rise to the high velocity sonic spike around 59 m.

As with underlying strata, the geophysical log signature of the Forest Marble shows considerable change as the formation is traced from the south coast to the area of the London Platform. Regional

correlation has not yet been effected and generally has been regarded as almost impossible. Towards the south, where the formation overlies mudstones, there is a distinctive sequence which is apparently correlatable over considerable distances as is illustrated by the comparable geophysical log signatures from boreholes over 25 km apart and separated by a major geological structure (Figs 33, 34). The repetitive upward changing motif is the response to the series of stacked sedimentary cycles described by Penn (1982, p.57) whereby, when fully developed, mudstones become increasingly silty upwards and pass into sandstones which are themselves overlain by shelly limestones or shell-beds. The shelly limestones or shell-beds give rise to the prominent sonic spikes and frequently lie with erosional contact on the hardened top of underlying beds as for example the Boueti Bed, the Digona Bed and a higher shell-bed at Seabarn Farm also characterized by *Goniorhynchia boueti* (Fig. 33). Careful comparison of core and log signature shows that the tip of the sonic spikes corresponds to the base of the shell-bed at, for example, the base of the Forest Marble at Seabarn Farm. This is because both the burrowed top of the underlying beds and the basal part of the Boueti Bed appear to be tougher and more calcareous, perhaps by diagenetic enhancement, than surrounding strata. The upper portion in which mudstones again are common, tends towards some of the character of the log signature of the lowest part (Fig. 33). Farther north and north-eastwards, the formation is not so readily subdivisible or correlateable, showing in places (Fig. 34) a monotonous, clay-dominated sequence; but in other places (and perhaps more commonly) mudstone-dominated successions, with characteristic higher gamma ray values and lower sonic velocity log responses, are confined to the top of the formation.

The log signature of the Cornbrash substantially repeats that of the Fuller's Earth Rock and Wattonensis Beds since it, too, comprises interbedded argillaceous limestones and shell-grit mudstones. Usually it is easily recognizable because prominent and widespread mudstones occur both above and below it. Generally the sonic velocity peak and low gamma ray values of the Cornbrash exceed those of the overlying Kellaways Sand and Rock of southern England (Figs 33, 34). North of the London Platform, however, the reverse is commonly true (Fig. 34). The formation often rests non-sequentially on Forest Marble, in places on a thin Forest Marble limestone which gives characteristic high sonic velocity peak and care has to be taken to exclude this limestone from the Cornbrash. Only where the Cornbrash is thick has it proved possible to determine the position of the boundary between Lower Cornbrash and the ubiquitous Upper Cornbrash from the log signature, so locating the base of the Callovian Stage. In thick sequences (e.g. Fig. 33) the Lower Cornbrash can usually be identified from the small-scale serrations of the otherwise uniformly low value, gamma ray curve in contrast to the large, more well defined, gamma ray peaks of the Upper Cornbrash. This difference reflects the generally more massive nature of the Lower Cornbrash in contrast to the interbedded, discrete limestone posts and commonly shell-detrital mudstones of the Upper Cornbrash.

North of the London Platform the log signature of Bathonian strata is dominated by an alternation of high and low gamma-ray values and sonic velocity which gives them a spikey log motif reflecting alternation of mudstones with thin siltstones, limestones and, in places, sandstones. This signature contrasts with the less erratic curves of the more homogeneous strata above and below (Fig. 34). The sandy beds at the base of the Upper Estuarine Beds give rise to low gamma ray values and high sonic velocity but pass upwards into mudstones, some of which are carbonaceous and yield high gamma ray values and markedly low sonic velocity. The overlying Blisworth Limestone (Fig. 34) yields prominent low gamma ray values and corresponding high sonic velocity levels which appear to represent the culmination of the upward changes in values measured in underlying strata. Over much of the East Midlands the log motif is typically bifid. It is suspected that this bifid peak may conceal internal as well as basal non-sequence and that the apparent downwards continuation of the log profile into the undated Lower Estuarine Beds does not necessarily imply that the latter are of similar age to the lower part of the Blisworth Limestone. The overlying Blisworth Clay comprises thin, interbedded mudstones, limestones, siltstones and in places sandstone of similar aspect to those found in the correlative Forest Marble to the south and gives rise to a similar, spikey geophysical log motif. The limestones of the Cornbrash are usually thin in this area and the formation gives rise to a prominent high sonic velocity spike and corresponding low gamma ray values which are commensurate with those of the Blisworth Limestone (Fig. 34).

Middle Jurassic strata show abundant evidence of cyclical sedimentation; often more than one hierarchical level of cycle is present. Such repetition is also displayed on geophysical log traces. Large-scale cycles, however, are difficult to prove within the Inferior Oolite owing to the lithological similarity of the predominantly limestone sequence, its

thinness and the presence of internal unconformities. The geological evidence suggest that three major cycles corresponding approximately to the Lower, Middle and Upper Inferior Oolite may be present and that each of these cycles may represent the deposits of three 'shallowing upwards' sequences following initial transgression. Four similar such sequences exist in the limestone facies of the Bathonian stage (Arkell 1956, p.27). As with the Inferior Oolite, non-sequence is prevalent and the limestone facies too thin to demonstrate satisfactorily the presence of such cycles in the corresponding geophysical log signatures. The southward passage of each of the limestone formations into the thicker clay sequence (Cave 1977, Penn *et al.* 1979) suggest that the Lower Fuller's Earth, the combined Fuller's Earth Rock and Upper Fuller's Earth, the Frome Clay and the Forest Marble constitute four similar cycles. Such major cycles are easily recognized by means of their geophysical log signatures. The Cornbrash may be considered as a basal facies to a similar cycle including Kellaways Clay and Kellaways Sand and Rock.

Upper Jurassic (including Callovian) log signatures

For convenience, the Callovian is here included in the Upper Jurassic. The English sequence is mainly argillaceous and contrasts markedly with the predominantly calcareous Middle Jurassic succession. Considerable variation in thickness occurs and is related to the general Jurassic depositional framework or to early Cretaceous erosion and overstep. In addition, lateral facies change is known; for example, Callovian strata in North Yorkshire are predominantly sandy (the Osgodby Formation) in contrast to their argillaceous Lower and Middle Oxford Clay correlatives farther south. It should be emphasized, however, that such changes are of relatively minor importance compared with the overall widespread regional similarity of the stratigraphical succession.

A major feature of these strata is the concordance in large parts of the sequence of lithology, ecostratigraphic succession and ammonite succession. This has led to exceedingly refined chronostratigraphic subdivision so that extremely thin sequences may be traced over hundreds of kilometres (e.g. Cox & Gallois 1981).

A second major feature is the way in which the regional upward changes in lithology may be referred to larger and smaller sedimentary cycles in each of which clay passes up through carbonate silt and sand to limestone. Thus the Callovian to Oxfordian passage from Oxford Clay to Corallian

Beds and the Kimmeridgian to Portlandian passage from Kimmeridge Clay through Portland Sand and Stone to Purbeck Beds have long been taken to represent major 'shallowing upwards' cycles each of approximately two stages in duration.

The significance of these major stratigraphical features to the present account is that the sequence readily lends itself to detailed regional correlation by means of geophysical log signatures. In the present account only the broader lithostratigraphical subdivisions are used. The Upper Cornbrash has been considered with the underlying part of the formation in the previous section.

Kellaways Beds

The geophysical log signature of the Kellaways Beds (e.g. Fig. 35) is highly distinctive. The basal part is almost invariably marked by a prominent gamma ray peak and corresponding low sonic velocity. The signature of the higher beds shows the overwhelmingly sandy nature of the Kellaways Sand and its upward passage from Kellaways Clay. The irregularity of the sonic signatures reflects the fact that the Kellaways Sand is in places poorly consolidated, while the prominent high velocity spikes are the response of levels which are tightly cemented with calcite.

It will be seen from Figs 35 to 37 that this signature is remarkably constant throughout England, despite different thicknesses of Kellaways Beds. It will be seen, too, that in those areas of Yorkshire where the Oxford Clay is absent through non-sequence, or the Kellaways Beds are overlain by sandy Middle and Upper Callovian strata, the overall Kellaways Beds signature is similar to that in southern England.

Oxford Clay

The gamma ray signature of the Oxford Clay (Figs 35, 36) is of a subdued serrated nature and shows, on average, higher values than those from subjacent and superjacent strata. In places there are prominent, high velocity sonic spikes which are almost as high as those registered from underlying or overlying strata. The most striking feature of the Oxford Clay log traces, however, is that their serrated nature (high frequency part of signal) is superimposed on larger scale, lower frequency variation.

These features reflect well the known lithological and stratigraphical variation within the Oxford Clay. Thus the high gamma ray values and subdued nature of the log traces indicate its predominantly argillaceous and monotonous character. Their serrated nature reflects the interbedded mudstones and muddy, calcite-siltstones. The highest gamma-ray values indicate the muddiest interbeds which are often carbonaceous and the highest sonic velocity spikes reflect beds of which the most calcareous are thin limestones. From this it will be seen that the largest scale variation is such that the high gamma ray values and lower sonic velocities predominating in the lower part of the sequence are readily identified with the mudstone-dominated, carbonaceous, Lower Oxford Clay sequence. The gradual upward change in motif reflects the increasing amount of carbonate silt and the presence of thin limestones characteristic of the Middle Oxford Clay. A minor sonic velocity spike and corresponding gamma low which breaks the gradual upward change is identified at Warlingham (Fig. 36) with the Acutistriatum–comptoni Bed and enables the Lower to Middle Oxford Clay boundary to be traced throughout southern England. The rapid upward change in motif marking the top of the Middle Oxford Clay may then be identified with the mudstones and shales characteristic of the basal part of the Upper Oxford Clay and marks the base of the Oxfordian Stage. As with the Lower and Middle subdivisions, the Upper Oxford Clay passes upwards into limier beds which in turn pass up to the Corallian Beds.

In the area of the East Midlands Shelf (Fig. 37) the log motif is similar but to the north the high gamma ray values of the lower part are missing, betokening major non-sequence within the Lower Oxford Clay. In this area, towards the margins of the basin, detailed correlation shows the Corallian Beds to rest with minor non-sequence on Upper Oxford Clay. In Yorkshire, the sandy equivalents (Langdale Beds and Hackness Rock) of the Middle Oxford Clay show a typical 'coarsening upwards' signature, while the Oxford Clay of Yorkshire has a signature similar to its southern counterpart (Fig. 37).

Corallian Beds and Ampthill Clay

The geophysical log signature of the Corallian Beds (clean, hard, shelly or oolitic limestones, in places quite thick) contrasts markedly with the underlying and overlying, predominantly mudstone sequences. It is characterized by very low gamma values and correspondingly high sonic velocities and high formation density. Preliminary investigation of the geophysical log stratigraphy suggests a twofold division of the Corallian Beds into a lower, more calcareous part (largely assigned to the Lower and Middle Oxfordian Substages) and an upper, more argillaceous part (largely assigned to the Upper Oxfordian Substage). Correlation in the upper part is rendered more difficult by the possible occurrence of Kim-

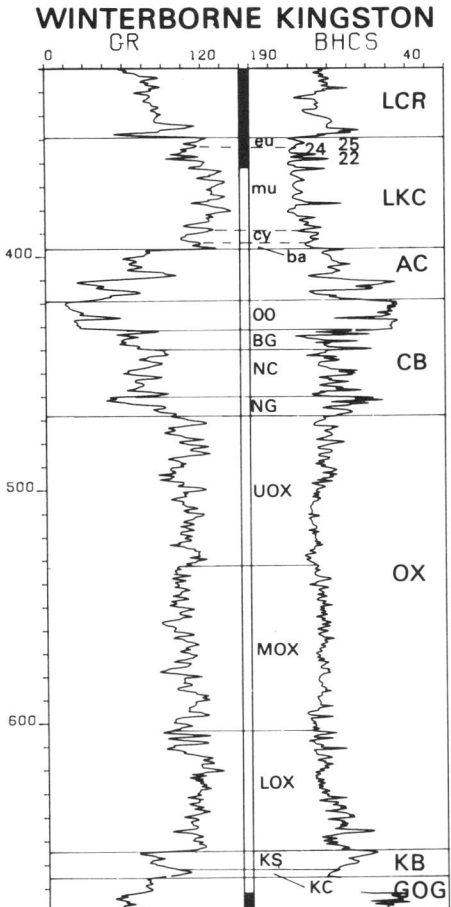

WINTERBORNE KINGSTON

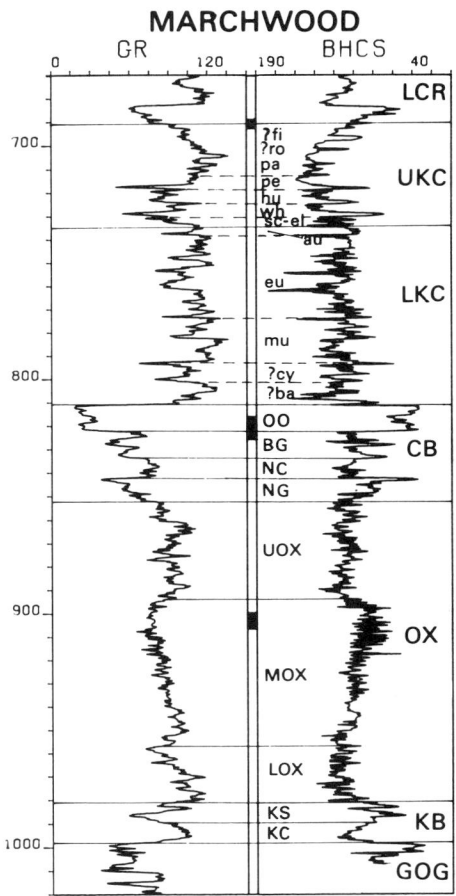

MARCHWOOD

FIG. 35. Upper Jurassic log signatures: Wessex Basin. GOG, Great Oolite Group; KB, Kellaways Beds; KC, Kellaways Clay; KS, Kellaways Sand; OX, Oxford Clay; LOX, MOX, UOX, Lower, Middle and Upper Oxford Clay respectively; CB, Corallian Beds; NG, Nothe Grit; NC, Nothe Clay; BG, Bencliff Grit; OO, Osmington Oolite; AC, Ampthill Clay; LKC, Lower Kimmeridge Clay including (ba) Baylei, (cy) Cymodoce, (mu) Mutabilis, (eu) Eudoxus and (au) Autissiodorensis zones and (22 to 25) bed numbers indicated; UKC, Upper Kimmeridge Clay including (el) Elegans, (sc) Scitulus, (wh) Wheatleyensis, (hu) Hudlestoni, (pe) Pectinatus, (pa) Pallasioides, (ro) Rotunda and (fi) Fittoni zones; PO, Portland Beds; PB, Purbeck Beds; GB, Gypsiferous Beds; JC, Jurassic–Cretaceous boundary; LCR, Lower Cretaceous; WB, Wealden Beds.

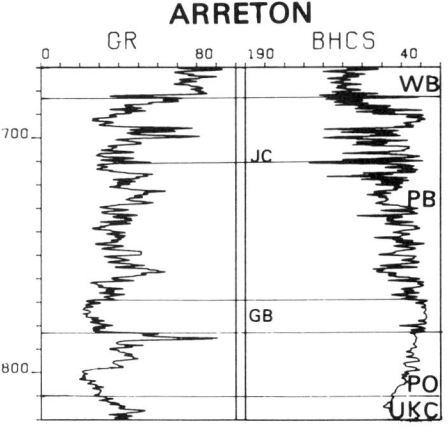

ARRETON

COLLENDEAN FARM   WARLINGHAM

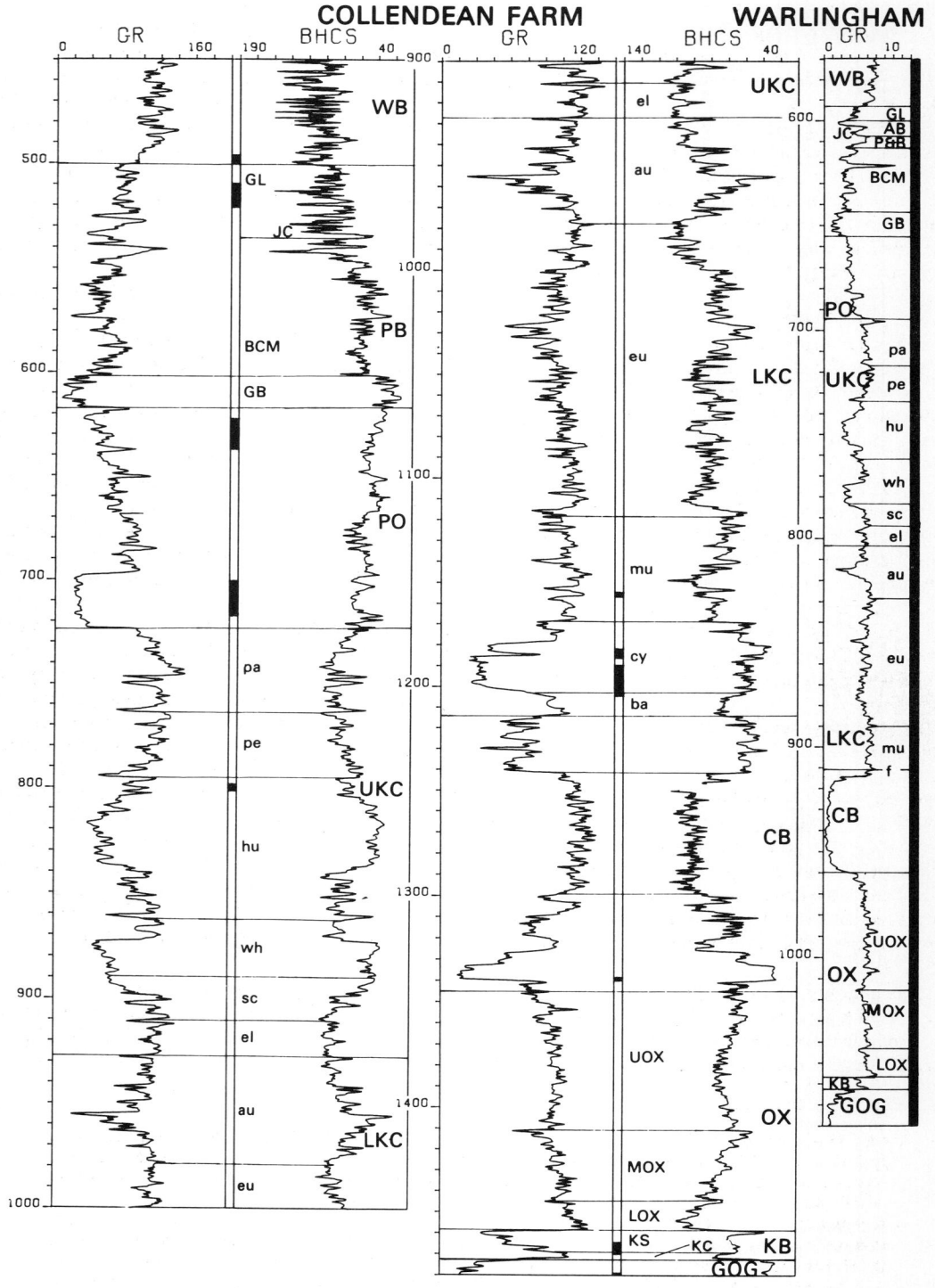

meridgian overstep, locally carrying a basal facies lithologically similar to the Corallian Beds of Upper Oxfordian age, but elsewhere carrying a predominantly clay facies (Kimmeridge Clay) which may rest locally on superficially similar clays of Oxfordian age.

In the Wessex area the lower part of the Corallian Beds has a characteristic signature as shown in Fig. 35. The twofold asymmetric signature is the response to two successive sedimentary cycles in each of which calcareous silt and sand increase in proportion as the sequence is climbed. The various parts of the signature represent, in upward succession, the Nothe Grit, Nothe Clay, Bencliff Grit and Osmington Oolite, or their correlatives of early- and mid-Oxfordian age.

Farther east, to the south of Warlingham in a thick Weald Basin sequence (Fig. 36), upper and lower limits of the Corallian Beds are readily identifiable despite the presence of a median mudstone member. Log correlation with neighbouring boreholes (some with core) shows that while the lower boundary is at a similar horizon to that elsewhere, the upper boundary is most probably near the top of the Cymodoce Zone of the Kimmeridgian Stage. In Fig. 36, the top of the Oxfordian, which in Dorset coincides with the top of the Corallian Beds, is at the top of the limestone unit around 1212 m depth. The upper boundary of the Corallian Beds so recognized is clearly diachronous.

In the axial part of the Weald Basin the Corallian Beds sequence is more argillaceous than that farther west (Wessex Basin) and the upper of the two sedimentary cycles is less prominent than the lower. At Warlingham, however, on the northern flank of the Weald Basin, the lower of the Corallian Beds cycles is sufficiently insignificant, and the beds sufficiently argillaceous, for them to have been regarded as part of the Oxford Clay and the upper cycle as part of a continuous mass of limestone (Fig. 36).

A similar log signature is found over much of the area of the East Midlands Shelf, although higher gamma ray values and lower sonic velocity usually occur because the Lower and Middle Oxfordian

West Walton Beds comprise mainly argillaceous, calcareous siltstones (Fig. 37).

In mid and north Yorkshire the Corallian Beds exhibit a signature comprising upward decreasing gamma ray values and increasing sonic velocity representing the upward passage from the calcareous siltstones and sandstones of the Lower Calcareous Grit to the cleaner limestones (Coralline Oolite) at the top of the Corallian Beds sequence (Fig. 37). Although the lower boundary of these Yorkshire Corallian Beds is at a similar horizon to that of Dorset and that suggested at Warlingham (Fig. 36), the top is at a much lower horizon, near the top of the Middle Oxfordian Substage.

The Ampthill Clay is characterized (Fig. 37) by a highly serrated geophysical log signature of moderate to high gamma ray values and generally low sonic velocity and formation density representing calcareous or silty mudstones. Superimposed on this signature are discrete levels of varying thickness, characterized by lower gamma ray values and higher sonic velocity. These represent beds of well-developed calcareous strata, some of which are thin limestones. They tend to occur at consistent levels (Fig. 37); indeed sufficient core material exists for the forty or so chronostratigraphic subdivisions of Gallois & Cox (1977) to be traced over much of the East Midlands. The upward increase in siltiness is particularly well shown by the formation density log and is of great importance as it is the local expression of the more general tendency towards more calcareous beds to occur near the top of the Oxfordian Stage. This trend coincides with the way in which these uppermost beds develop internal disconformities towards the south-eastern edge of the Eastern England Basin. These minor breaks herald the subsequent development of the basal Kimmeridgian unconformity where, in that area, the Kimmeridge Clay oversteps the highest Ampthill Clay (Gallois & Cox 1977, p.216).

It is not clear whether the exceptionally prominent break in the log signature marking the top of the Corallian Beds at Hunmanby (Fig. 37) is due to Kimmeridge Clay overstepping Ampthill Clay or

---

FIG. 36. Upper Jurassic log signatures: Wessex Basin. GOG, Great Oolite Group; KB, Kellaways Beds; KC, Kellaways Clay; KS, Kellaways Sand; OX, Oxford Clay; LOX, MOX, UOX, Lower, Middle and Upper Oxford Clay respectively; CB, Corallian Beds; f, fault; LKC, Lower Kimmeridge Clay including (ba) Baylei, (cy) Cymodoce, (mu) Mutabilis, (eu) Eudoxus and (au) Autissiodorensis zones; UKC, Upper Kimmeridge Clay including (el) Elegans, (sc) Scitulus, (wh) Wheatleyensis, (hu) Hudlestoni, (pe) Pectinatus and (pa) Pallasioides zones; PO, Portland Beds; PB, Purbeck Beds; GB, Gypsiferous Beds; BCM, Broadoak Calcareous Member; P & B, Plant and Bone Beds; JC, Jurassic–Cretaceous boundary; AB, Arenaceous Beds; GL, Greys Limestone; WB, Wealden Beds. Gamma ray log at Warlingham in counts/s.

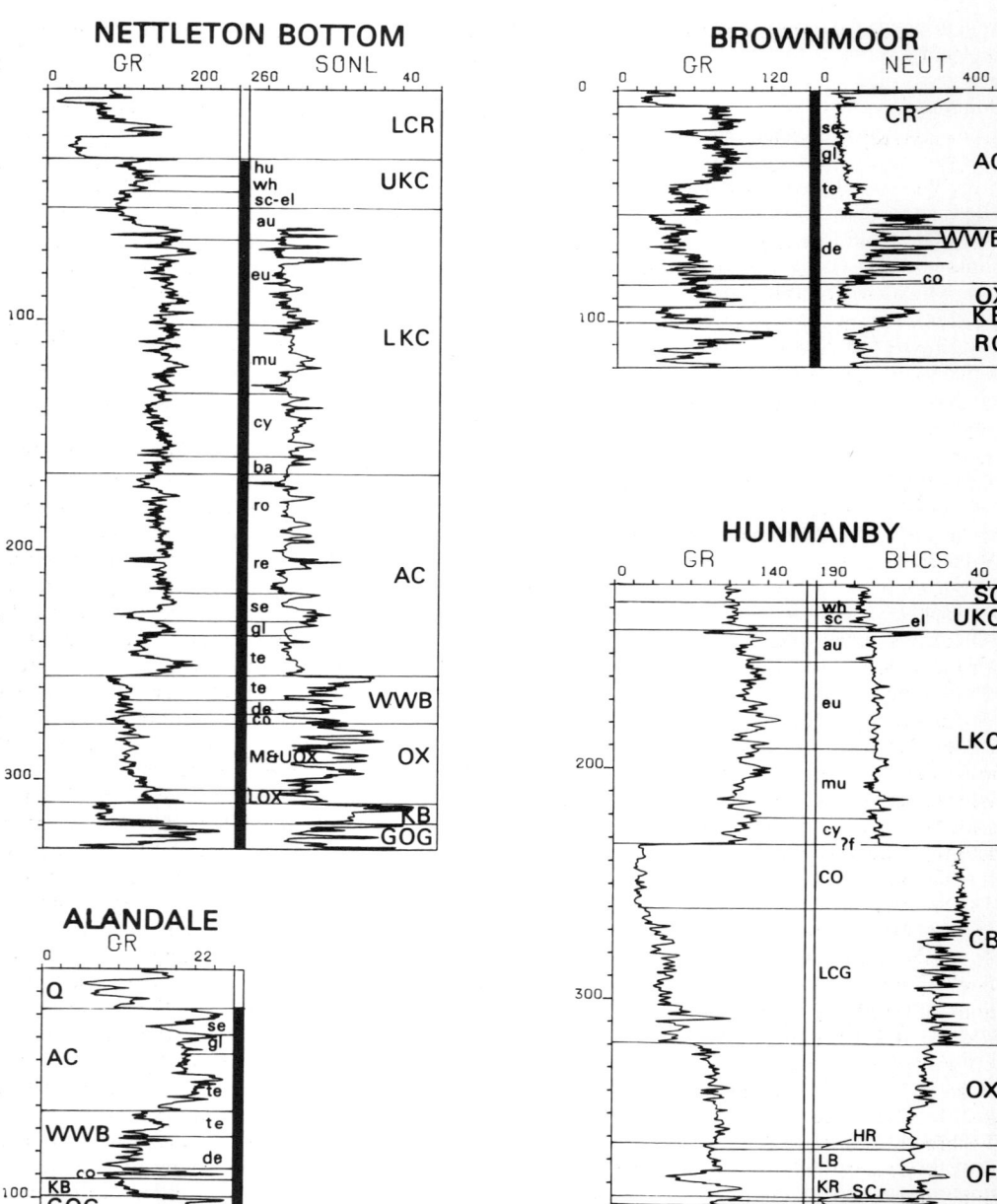

FIG. 37. Upper Jurassic log signatures: East Midlands Shelf and Yorkshire. GOG, Great Oolite Group; RC, Ravenscar Group; CBr, Cornbrash; SCr, Shales of the Cornbrash; KB, Kellaways Beds; OF, Osgodby Formation; KR, Kellaways Rock; LB, Langdale Beds; HR, Hackness Rock; OX, Oxford Clay; LOX, MOX, UOX, Lower, Middle and Upper Oxford Clay respectively; CB, Corallian Beds; LCG, Lower Calcareous Grit; CO, Coralline Oolite; WWB, West Walton Beds including (co) Cordatum, (de) Densiplicatum and (te) Tenuiserratum zones; AC, Ampthill Clay including (gl) Glosense, (se) Serratum, (re) Regulare and (ro) Rosenkrantzi zones; LKC, Lower Kimmeridge Clay including (ba) Baylei, (cy) Cymodoce, (mu) Mutabilis, (eu) Eudoxus and (au) Autissiodorensis zones; UKC, Upper Kimmeridge Clay including (el) Elegans, (sc) Scitulus, (wh) Wheatleyensis and (hu) Hudlestoni zones; CR, Cretaceous; LCR, Lower Cretaceous; SC, Speeton Clay; Q, Quaternary. Gamma ray logs at Nettleton Bottom and Alandale, and both logs at Brown Moor, in counts/s.

whether the undoubted absence of Ampthill Clay is due to faulting.

Kimmeridge Clay

Fundamental work concerning the lithology of the Kimmeridge Clay and its geophysical log response was carried out by Gallois (1973), who studied its variation in shallow-cored boreholes in the Wash area. He correlated lithological variation with gamma ray signature and subsequent study with Dr B. M. Cox established 48 subdivisions of the formation which were traced from Yorkshire to the type area of the Dorset coast (Gallois & Cox 1974, 1976, Gallois 1979, Cox & Gallois 1981). The boundaries of these units are sensibly parallel to those of the ammonite zonal stratigraphy and there is every reason to regard them as chronostratigraphic subdivisions. Subsequent work has enabled most of them to be traced throughout the subsurface of the East Midlands by means of their geophysical log signature (Penn, Cox & Gallois, in preparation). Is is thus possible (in regions away from structural complexity or facies complication) to elucidate the stratigraphy of the Kimmeridge Clay to less than zonal level with a minimum of palaeontological assistance.

The Kimmeridge Clay is composed of a complex sequence of small-scale sedimentary rhythms. In the lower part these comprise thin siltstones or silty mudstones, resting on an erosion surface, and overlain by medium or dark grey, shelly, fissile mudstones which become lighter coloured, more calcareous and less fissile upwards. In the remainder of the Kimmeridge Clay these rhythms consist of bituminous shales (fissile, brownish black, shelly and phosphatic mudstones) grading upwards by increasing carbonate content through medium and dark grey mudstones into pale grey, calcareous mudstones and argillaceous limestones. Superimposed on these small-scale rhythms are larger scale changes from more or less calcareous to more or less bituminous beds.

The large-scale variations in calcimetry and also in kerogen content combined with the smaller scale rhythms together produce a very distinctive geophysical log pattern. It has a highly serrated sonic and formation density log signature. Gamma ray values are generally high, and show complementary but more subdued variation to those of the sonic log. Bituminous beds are characterized by high gamma ray, low sonic velocity and very low density values. Thus the typical Kimmeridge Clay rhythm gives rise to a log response in which high gamma ray values and low sonic velocity characterize the shaley, lower part of each rhythm. Where this part is bituminous very low formation density

values are also recorded but, as in the lower parts of the Kimmeridge Clay, a basal sandstone may give rise to a high velocity sonic spike and low gamma ray values although the latter values may be higher if phosphatized material is present. The gamma ray values gradually decrease through the rhythm, while the sonic velocity and density values increase complementarily. At the top of the rhythm the calcareous beds give rise to the lowest gamma-ray values and highest sonic velocities. When these beds become typical porcellanous limestones, pronounced sonic and density 'spikes' are registered. Log signatures of the Kimmeridge Clay rhythms are shown in Figs 35 to 37.

Having identified the individual components of the log signature of the Kimmeridge Clay in full sequences well known from coring, it is possible to make some generalizations about the log curve. First of all the prominent high velocity sonic spikes and low gamma ray values representing hard beds such as coccolith-rich limestones, though laterally widespread, may be difficult to identify individually because intervening gamma ray peaks are less persistent, reflecting the less persistent nature of the oil–shale seams. Consequently it is more useful, when examining a log signature that has poor palaeontological control, to smooth the log curve by eye. This results in the clustering of peaks representing the more limey levels.

The Baylei and Cymodoce zones are difficult to detect because in places they carry sufficient silt, sand or lime to be confused with the underlying Corallian Beds. The Mutabilis Zone, however, generally shows upward decreasing gamma ray values. The smoothed gamma ray curve of the Eudoxus Zone shows, in a subdued way, a symmetrical curve similar to that developed in the overlying Autissoidorensis, Wheatleyensis and Hudlestoni zones; within these last three zones it is sometimes possible to recognize three peaks corresponding to the development of limestone beds. Above this level the gradual upward increase in gamma ray values through the top Hudlestoni and Pectinatus zones is followed by gradual upward decrease through the Pallasioides, Rotunda and Fittoni zones as they pass up to Portland Beds. One of the corollaries of this approach is that if the Callovian and Oxfordian stages be regarded as basically having upward decreasing gamma ray motifs, with the latter more strongly developed than the former, then the corresponding upper limit of the Kimmeridgian Stage should be taken around the top of the Pectinatus Zone, i.e. around the top of the major upward decreasing gamma ray motif of the Kimmeridge Clay. It is noteworthy that it is exactly at this level that Casey (1967) suggested the top of the Kimmeridgian Stage should be taken.

Difficulty may be encountered in subdividing the Kimmeridge Clay in areas of stratigraphical complexity. At Winterborne Kingston, for example, it is very thin despite having accumulated in a Jurassic trough. Fortunately the Mutabilis/Eudoxus zonal boundary was determined from core material. The Mutabilis Zone has a typical downwards increasing gamma ray signature with a median, silty level, probably Bed 17, being prominent (Fig. 35).

A thin sequence was also encountered at Marchwood (Fig. 35) but here no core material was available and the correlation is more difficult because of the likely break in the sequence at the top of the Corallian Beds. However, the logs indicate that the smooth downward increase followed by decrease in gamma ray values represents the usual Kimmeridgian sequence from the Rotunda Zone down to the top of the Hudlestoni Zone. Below here the trifurcate low gamma ray values from 715–740 m represent the Hudlestoni Zone to Autissiodorensis Zone sequence and the underlying low gamma ray values around 750–760 m represent the typical Eudoxus Zone motif. Gamma ray values show an overall decrease to about 790 m as would be expected in the Mutabilis Zone, in which case the prominent low gamma ray values at 790–800 m above the gamma ray peak at 805 m indicate the typical Cymodoce Zone to Baylei Zone sequence. Such an interpretation implies that the Upper Oxfordian is extremely condensed or missing. Should this prove not to be the case then stratigraphic condensation must occur within the Kimmeridge Clay.

Log signatures from the Kimmeridge Clay of Yorkshire can generally be satisfactorily correlated with those of the East Midlands (Fig. 37).

Portland Beds

A full Portland Beds sequence has rarely been drilled and geophysically logged. This is because thick developments of these strata are confined to an area south of the London Platform. In the type area south of the Abbotsbury–Isle of Wight line they are for the most part, structurally high and well exposed. North of here they have been removed over Jurassic 'highs' by pre-early Cretaceous erosion. Still farther north, in the Weald Basin, the Portland Stone is absent or poorly developed owing to the early onset of Purbeck Beds sedimentation or by sub-Purbeck Beds disconformity.

The geophysical log signature of the Portland Beds (Fig. 36) is characterized by upward decreasing gamma ray values and increasing sonic velocity as silty mudstones pass up through muddy siltstones to fine-to-medium-grained sandstones. The signature may be 'spikey' because of the occurrence of sporadic levels well cemented by dolomite or calcite or even of thin limestones. Thick units of low gamma ray values may occur and may be associated with low sonic velocity, reflecting the fact that substantial parts of Portland Sand are poorly cemented (Fig. 36). Commonly, however, the gamma ray values themselves are not particularly low since the sands are frequently argillaceous or glauconitic.

Purbeck Beds

The whole of the Purbeck Beds are discussed here, though the upper levels are of Cretaceous age. Their geophysical log signature is not well known from areas other than the Weald. Here the beds come to outcrop as faulted inliers and are known from surface mapping, shallow drilling and mining to exploit their basal evaporites. Elsewhere they are known from hydrocarbon exploration wells and BGS cored boreholes.

Their signature is distinctive (Figs 35, 36). Low gamma ray values, high sonic velocities and very high formation densities mark the interbedded gypsum–anhydrite, thin limestones and mudstones of the basal Gypsiferous Beds of the Purbeck Beds of the Weald (Fig. 36). Above, the log motif takes on its usual highly serrated nature marking the alternating thin bedded, micritic limestones and calcareous mudstone of the Broadoak Calcareous Member (Blues Limestone) (Fig. 36). Higher up, the sequence is hard to characterize. There is a marked increase in gamma ray values being the log response to the mudstones and siltstones of the Plant and Bone Beds and the lower, silty part of the Arenaceous Beds. Within this part of the sequence minor low gamma ray values correspond to the shelly mudstones and thin limestones of the Cinder Bed, marking the Jurassic/Cretaceous boundary (Figs 35, 36). The upper, sandy part of the Arenaceous Beds gives rise to prominent low gamma ray values and, with the return to the typical serrated signature of the overlying interbedded thin limestones and mudstones of the Grey Limestones, imparts a minor asymmetry to the upper part of the Purbeck Beds geophysical log signature. There is a general upwards increase in gamma ray values (and decrease in sonic velocity) as the Purbeck Beds pass up to Wealden Beds, particularly in the axial part of the Weald Basin where the basal Wealden Beds are predominantly muddy.

Lower Cretaceous log signatures

Lower Cretaceous strata show considerable vertical and lateral variation in thickness and facies; generally the sequence is thickest and fullest towards the basin centres. In the Weald Basin and the area south of the St Valery–Bembridge line, thick Ryazanian to Barremian non-marine deposits are usually conformable on Jurassic strata; they thin to the basin margins where they are overlapped and overstepped by the succeeding marine Aptian–Albian sediments. The latter rest on Jurassic or older strata on these marginal 'highs'. The most prominent high is the London Platform, over large areas of which pre-Albian strata are absent. North of the platform thin, condensed Ryazanian to Barremian marine strata are overlain disconformably by an attenuated Aptian–Albian sequence: the succession forms the thin south-western margin of the southern North Sea Basin. Over the Market Weighton High, Albian strata come to rest on Permo-Triassic rocks, while to the north Red Chalk rests on thick Speeton Clay.

The stratigraphical terminology of Rawson *et al.* (1978) is followed here. For convenience, the whole of the Spilsby Sandstone and Sandringham Sands is discussed in this section, though the lower beds are of latest Jurassic age.

Geophysical logs are available for the whole Lower Cretaceous sequence, from wells drilled for hydrocarbon and geothermal purposes. Of considerable importance, however, are shallow boreholes drilled by the British Geological Survey, mainly for stratigraphical purposes. Though containing less varied logging suites, usually these are cored throughout and enable calibration of the log signature with the rock succession, particularly in the condensed sequences to the north of the London Platform.

The Wealden Beds

The thick Wealden Beds comprise a lower, sandstone-dominated sequence and an upper, more argillaceous one. In the Weald Basin these are called the Hastings Beds (Ryazanian to Valanginian) and the Weald Clay (Hauterivian and Barremian) respectively. The Hastings Beds embrace three major cycles in each of which claystones and mudstones coarsen upwards into cross-bedded sandstones. Above these there is a passage through cross-laminated siltstones to mudstones of the next cycle.

Although characteristically 'ratty', the gamma ray signature of the Ashdown Beds (Fig. 38) shows an overall upward decrease in values in which the higher value 'peaks' become less and less prominent. This reflects the gradual upwards passage from the shales and mudstones with subordinate siltstones of the Covehurst Bay Member (Fairlight Clay) through a silty sequence to the predominantly thickly bedded sandstones of the upper part of the Ashdown Sands. Higher values of similarly serrated gamma ray profiles occur between the low values associated with the sandstone bodies and represent interbedded siltstones. The sonic log shows low velocity mudstones, especially to the base, with moderate velocity peaks reflecting the poorly consolidated nature of both mudstones and interbedded siltstones. The higher, sandier part of the sequence has a more monotonous signature of moderate sonic velocity indicating that the sandstones and siltstones are not particularly well consolidated. Similar gamma ray and sonic log motifs as found in the Ashdown Sands occur in the overlying Wadhurst Clay and Lower Tunbridge Wells Sand succession as well as in the Grinstead Clay and Upper Tunbridge Wells Sand sequence. The Weald Clay in the centre of the basin is poorly known; generally high gamma values and a moderately serrated motif are found.

The Warlingham Borehole (Fig. 38) may be more representative of a basin margin succession. At Warlingham the same distinction is encountered between the serrated lower gamma ray signature of the Hastings Beds and the less serrated higher value signature of the Weald Clay. Unfortunately, however, the casing point around 462 m shifts the gamma ray curve leftwards and lessens the effect of the contrast. The Hastings Beds gamma ray signature has much less prominent low values denoting the lesser frequency of sandstone bodies as compared with the corresponding beds towards the basin axis (Fig. 38). Nevertheless the alternating sandstone/mudstone sequence of the Ashdown Beds, Wadhurst Clay, Lower Tunbridge Wells Sand, Grinstead Clay and Upper Tunbridge Wells Sand is still obvious in this thinner succession. This alternation and the higher clay content of the Warlingham sequence is well brought out in the SP curve which shows marked negative separation in the sandier horizons and positive separation approaching that of the Weald Clay in mudstones. The marked horizontal deflection associated with the Wadhurst Clay, however, cannot be achieved in a normal SP curve and it is likely that it, with those at 340 m, 350 m and 433 m, reflects interference from local industrial sources (Worssam & Ivimey-Cook 1971, p. 112).

The monotonous, yet finely serrated, gamma ray curve of higher values reflects the predominantly mudstone sequence with interlaminated siltstones and very thin limestones and ironstones characteristic of the Weald Clay (Fig. 38). Lower gamma ray

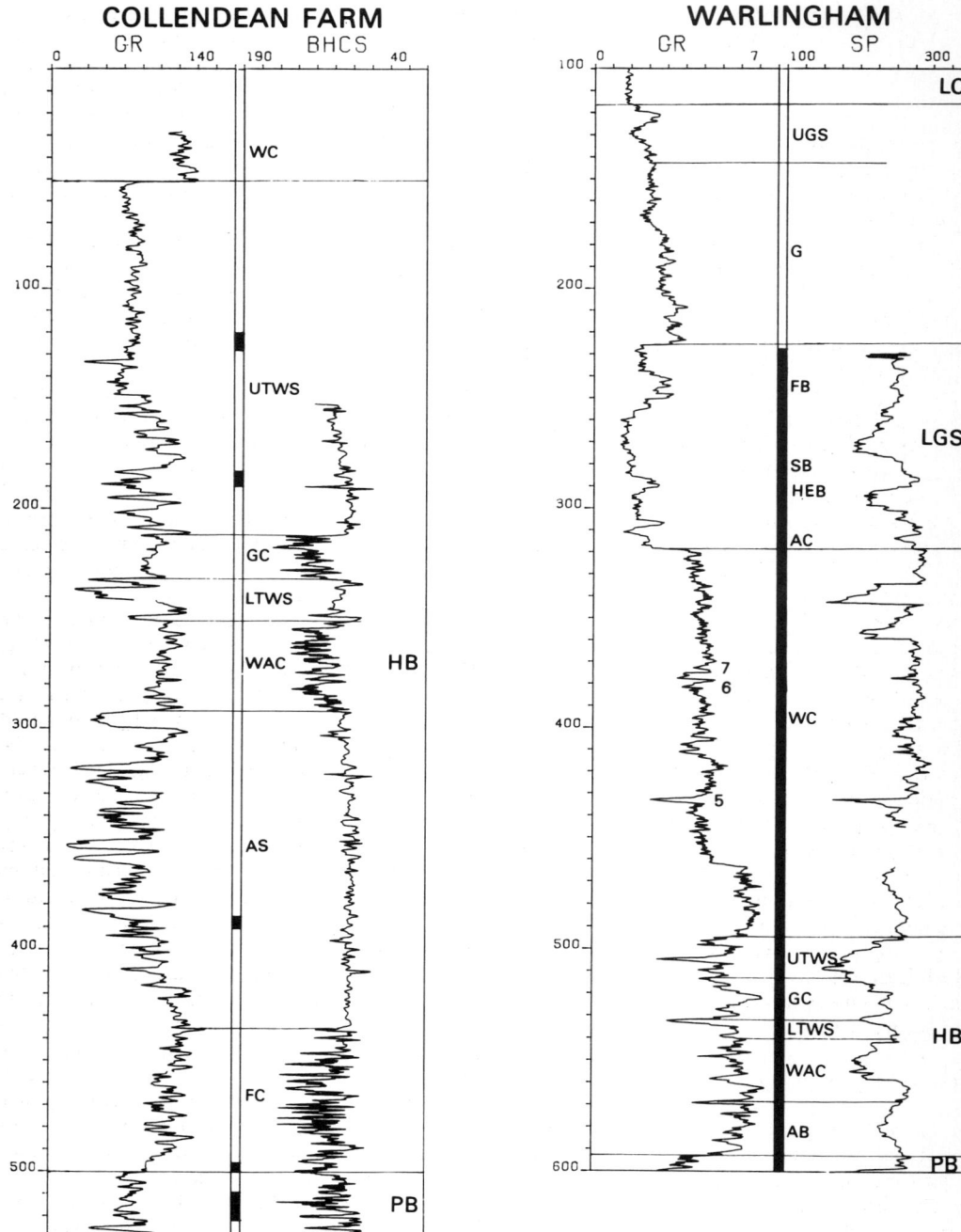

Fig. 38. Lower Cretaceous log signatures: Weald Basin. PB, Purbeck Beds; HB, Hastings Beds; FC, Fairlight Clay; AS, Ashdown Sands; AB, Ashdown Beds; WAC, Wadhurst Clay; LTWS, Lower Tunbridge Wells Sand; GC, Grinstead Clay; UTWS, Upper Tunbridge Wells Sand; WC, Weald Clay; LGS, Lower Greensand; AC, Atherfield Clay; HEB, Hythe Beds; SB, Sandgate Beds; FB, Folkestone Beds; G, Gault; UGS, Upper Greensand; LC, Lower Chalk.

values, however, correspond to the sandstones and siltstones which have been identified in cores as Topley's Beds 7, 6 and 5 (Worssam & Ivimey-Cook 1971).

The Aptian–Albian sequence of southern England

The tripartite division of the marine Aptian–Albian beds into Lower Greensand, Gault and Upper Greensand, is reflected by the gamma ray signatures: low values characterize the glauconitic sandstones of the Lower and Upper Greensand and higher values, the intervening mudstones of the Gault. Within the Lower Greensand, however, higher gamma ray values occur reflecting intervening siltstones and mudstones some of which are sufficiently prominent as to merit lithostratigraphical subdivision (Fig. 38).

The base of the Gault is marked by a sharp rise in gamma ray values reflecting both the sharp contact with the Lower Greensand and its often phosphatic nature. Lower gamma ray values occur within the Gault marking the occurrence of silty levels and high values mark the occurrence of phosphatized levels. There is a gradual upwards decrease in gamma ray values marking the upwards passage into the Upper Greensand. Although the lithological transition from Gault to Upper Greensand is diachronous these gamma log features of the Aptian–Albian succession are generally constant in southern England, even where the Lower Greensand comes to rest on older strata. For example at Marchwood (Fig. 39) the tripartite subdivision is still evident within a much thinner sequence. Here the Lower Greensand is less silty to the top and has well cemented horizons reflected in peaks of high sonic velocity. The sharp base to the Gault and its gradual upward passage to the Upper Greensand is seen in both gamma ray and sonic log curves. Here the sonic log shows higher velocities as well as marked high velocity peaks in the upper part, which is well cemented and contains frequent posts of sandy limestone.

Lower Cretaceous rocks north of the London Platform

On the northern flank of the London Platform (in Norfolk), the glauconitic Sandringham Sands rest with marked unconformity on Kimmeridge Clay. The former yields low gamma ray values with minor peaks which represent interbedded clayey sands (Fig. 39). Prominent high value spikes occur at the base and at certain higher levels and reflect the occurrence of phosphatic nodule beds and phosphatized erosion surfaces. The sonic log (Fig. 39) trace is correspondingly mildly serrated and is of moderate to low sonic velocity reflecting the mildly cohesive nature of the deposit. High velocity spikes representing better cemented levels occur though some correspond to the phosphatized levels marked by gamma ray peaks. The phosphatized levels are of great stratigraphical importance in these condensed sands and have been used to subdivide them. Indeed one marks the Jurassic/Cretaceous boundary, the underlying beds being of middle and late Volgian age and the upper beds Ryazanian to Valanginian.

Overlying the Sandringham Sands are Hauterivian and Barremian clays with basal sands and silts, the Dersingham Beds. These register higher gamma ray values than the underlying sandstones though the interbedded silts give rise to prominent low gamma ray values.

The succeeding Albian strata comprise the Carstone and overlying Red Chalk. The glauconitic sandstones of the former are marked by low gamma ray values, in contrast to the underlying strata, but interbedded clays give rise to a serrated signature. The Red Chalk has a characteristic signature which shows an overall upward decrease in gamma ray values as the terrigenous content of the limestone decreases upwards.

To the north, Lower Cretaceous strata of the East Midlands Shelf exhibit similar log motifs (e.g. Kirby *et al.* 1985) (Fig. 39). The Spilsby Sandstone, like the Sandringham Sands, comprises hard and soft sandstones, very thin clays and phosphatized beds, and has a similar log signature. Usually the Mid-Spilsby Nodule Bed, marking the Jurassic/Cretaceous boundary, can be recognized. Again these sandstones are succeeded by a dominantly mudstone sequence of higher gamma ray and lower sonic velocities. However, the sandstone/mudstone boundary is of Ryazanian age here as opposed to Valanginian in Norfolk. The mudstones and ferruginous clays and sands range from latest Ryazanian to Aptian in age and are unconformably overlain by a thin Carstone which is usually marked by a high velocity sonic 'spike' and corresponding low gamma ray value. Again, though thin, the Red Chalk is characterized by upward decreasing gamma ray values and upward increasing sonic velocity reflecting its upward decrease in terrigenous detritus and concomitant increase in lime content. The upward change in log signature is frequently rapid as the Red Chalk is usually very thin and condensed but the signature exhibits a slower upward change where the formation is thicker.

North of the Market Weighton area Lower Cretaceous strata comprise the thick clays and thin nodule beds of the Speeton Clay, in which all Lower Cretaceous stages have been proved, over-

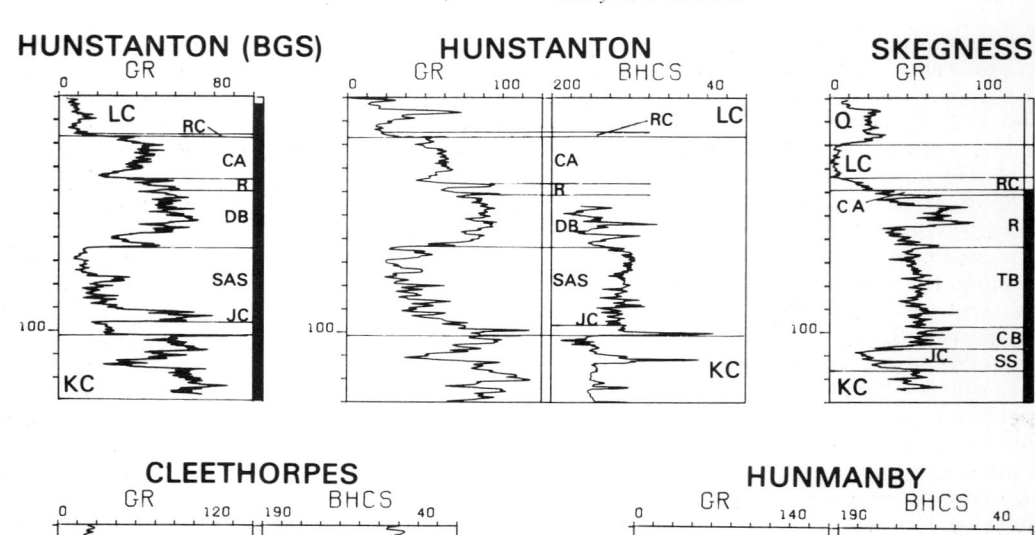

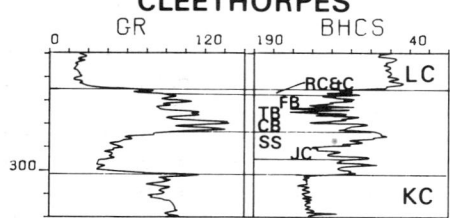

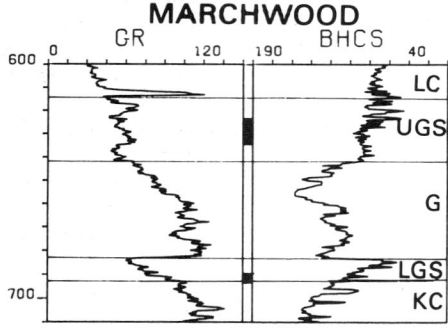

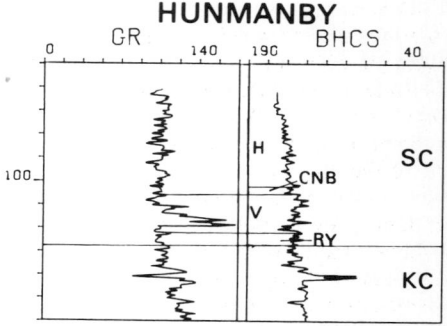

FIG. 39. Lower Cretaceous log signatures. KC, Kimmeridge Clay; JC, Jurassic–Cretaceous boundary; SAS, Sandringham Sands; SS, Spilsby Sandstone; SC, Speeton Clay; CB, Claxby Beds; TB, Tealby Beds; FB, Fulletby Beds; DB, Dersingham Beds; R. Roach; LGS, Lower Greensand; CA, Carstone; G, Gault; RY, Ryazanian; V, Valanginian; H, Hauterivian; CNB, Compound Nodular Bed; UGS, Upper Greensand; LC, Lower Chalk; RC, Red Chalk; RC & C, Red Chalk and Carstone; Q, Quaternary. Gamma ray log at Hunstanton (BGS) and Skegness in counts/s.

lain by typical Red Chalk. The Speeton Clay rests with marked disconformity on Kimmeridge Clay. At Hunmanby detailed log correlation shows the highest Kimmeridge Clay to lie around the base of the Hudlestoni Zone. Speeton Clay log signatures show overall low sonic velocity reflecting the overall softness of the formation. High and low gamma ray values and sonic velocity occur reflecting the interbedded, dark, commonly glauconitic clays and paler grey, harder, calcareous or nodular beds. Comparison with cored sequences shows that log signatures of Ryazanian to lower Hauterivian strata can be recognized throughout Yorkshire. Noteworthy is the characteristic signature of the

Valanginian strata, which show prominent gamma ray peaks with overall upward decreasing values.

Upper Cretaceous log signatures

The familiar English Chalk is in the main a white, very pure, fine-grained limestone composed of coccolith fragments. As has often been remarked, however, this homogeneity conceals considerable lithological variation caused by varying admixtures of clay, silt or even sand, often glauconitic, as well as much shelly fossil debris. It is, moreover, extremely susceptible to diagenetic alteration, including the formation of more or less widespread, condensed, hardground sequences, some of which may be phosphatized or glauconitized. In parts of the succession, are prominent sheets and irregular nodules of flint which, though commonly parallel to bedding, may well be diagenetic.

The combination of such lithological variants has resulted in a multiplicity of local stratigraphical names and there is no real consensus as to which names are regionally apt. In particular there is a long established distinction between the more massive and softer chalks of the Southern Province and the harder, thin-bedded chalks in the Northern Province (Norfolk to Yorkshire). Although there is lateral passage between them, each province belongs to a basin with a different depositional and post-depositional history; a fact which is reflected in their faunas and which makes biostratigraphical correlation difficult.

Despite these differences there is an overall similarity in the Chalk sequence that is brought out by comparison of geophysical log signatures from each of the provinces and which gives rise to the prospect of a detailed countrywide geophysical log correlation eventually being established. This would thus have considerable impact on stabilizing stratigraphical nomenclature. At present, the nomenclature of Rawson *et al.* (1978) is followed: this retains the familiar Lower, Middle and Upper Chalk divisions in both provinces. There are problems about the precise location of the stratigraphical limits of these subdivisions (Wood & Smith 1978, pp. 265–269) but this does not necessarily mean that such widespread overall subdivisions do not exist, merely that greater effort should be made in regional correlation. Nor does it necessarily replace the need for local names as for example those erected by Wood & Smith (1978) and Mortimore (1983).

Geophysical logs have been obtained from many of the numerous shallow boreholes drilled to test and develop the Chalk as an aquifer. Initially these were single point resistance logs, often with wide electrode spacing, but subsequently continuous logging became accepted practice. In later years, more varied log suites (including gamma ray, density, neutron and sonic logs) have become available particularly because of drilling by BGS and the hydrocarbon industry for scientific and exploration purposes. The use of such logs in stratigraphical correlation stems from work in the London area by Gray (1958, 1965) who calibrated log signatures against continuously cored sequences. This work was extended by his colleagues who have been able to trace marker beds by means of their log signatures as far north as Yorkshire (Murray 1982a,b). The most recent work has correlated log signatures with local successions in great detail in Lincolnshire and South Humberside (Barker *et al.* 1984). In the Weald and Hampshire Basin, Mortimore (in preparation) has incorporated geophysical log signatures in his proposed new classification of the Chalk of the Sussex area.

The gamma ray signature of the Chalk (Figs 40, 41) is highly distinctive, being characterized by exceptionally low (often less than 10) API units about which there are minor oscillations. The monotony of the curve is broken by rare sharply defined peaks of higher API units corresponding to the more prominent marl bands (or the more sporadic phosphatized chalks) and by the consistently higher units registered in the more marly Lower Chalk. Better definition may be seen in the sonic log and to a certain extent in the resistivity logs. There is generally an overall upward decrease in sonic velocity corresponding to the general tendency of softer chalks to occur towards the top of the Chalk. The curve is moderately serrated, however, reflecting the alternation of hard and soft chalks with interbedded marls and flints. Although there is an overall similarity of sonic log motif throughout the Chalk, there appears to be a major distinction in absolute values of interval transit-time between the two provinces: much faster velocities are usually recorded from the Northern Province, except in the uppermost beds where they approach the slower velocities of the Southern Province. The difference appears to reflect the well-known distinction between the hard limestone chalks of the Northern Province and the softer 'earthy' chalks of the Southern Province.

The prominent marl levels can usually be detected by the coincidence of higher gamma ray values and lower sonic velocities. In flint-free Chalk the hard chalks can usually be distinguished from the softer chalks by the higher, often markedly so, sonic velocity peaks of the former. Where such hard chalks are phosphatized gamma ray values may also increase but where flints are known to be present they cannot usually be separated from unphosphatized hard chalks without recourse to other data, particularly since the density/neutron tool has been rarely run in onshore boreholes. In general the resistivity curves follow that of the sonic log although resolution is poorer and they cannot be used above the water table or where saline invasion has taken place. On the other hand the marly levels tend to be particularly well marked being characterized by prominent sharp peaks of low resistivity values.

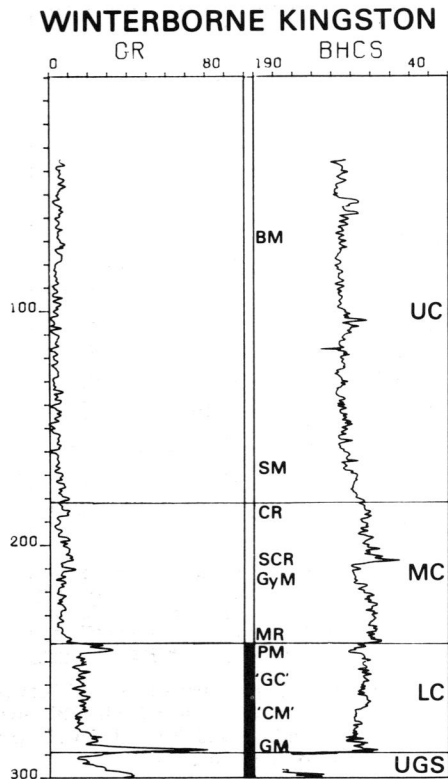

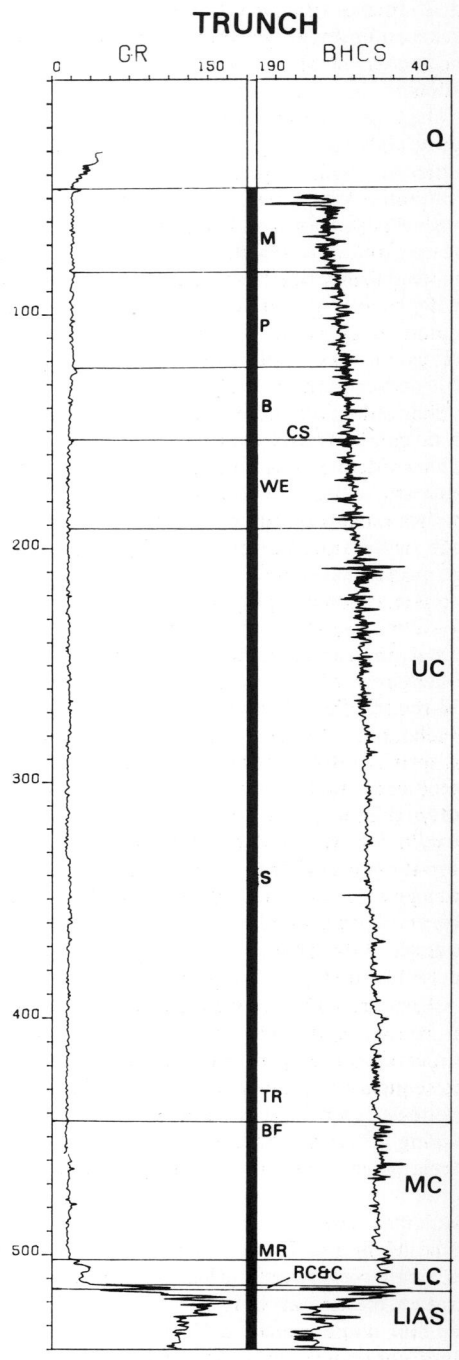

FIG. 40. Upper Cretaceous log signatures.
UGS, Upper Greensand; LC, Lower Chalk;
RC & C, Red Chalk and Carstone; GM,
Glauconitic Marl; 'CM', 'Chalk Marl'; 'GC',
Grey Chalk; PM, Plenus Marl; MC, Middle
Chalk; MR, Melbourn Rock; GyM, Glynde
Marl; SCR, Spurious Chalk Rock; CR, Chalk
Rock equivalent; BF, Brandon Flints; UC,
Upper Chalk; TR, Top Rock; SM, Shoreham
Marl; S, Seaford Marl; BM, Brighton Marl;
WE, Weybourne and Eaton chalks; CS,
Catton Sponge Beds; B, Beeston Chalk; P,
Paramoudra Chalk; M, Maastrichtian Chalk;
Q, Quaternary.

The Lower Chalk

Although the broad lithological subdivisions of the Southern Province Lower Chalk and their corresponding log signatures (Fig. 40) are usually easily recognizable, their precise boundaries may be diachronous. The base of the Lower Chalk is identified by the prominent gamma ray peak which corresponds to the basal, glauconite siltstone or sandstone which frequently contains reworked or phosphatized fragments of underlying strata. This hard bed is also commonly marked by a sonic peak. The upward diminishing clay and silt content as the overlying Chalk Marl passes up into Grey Chalk is usually clearly brought out by the steady decrease in gamma ray values. A similar gradual upward increase in sonic velocity occurs but the signature of the latter is moderately serrated reflecting the interbedded chalk and marl. Individual marls and chalks can be traced over considerable areas and may provide the basis for a refined subdivision, but systematic correlation and integration with the detailed biostratigraphic sequence has yet to be made. The Plenus Marls are marked by a moderate but prominent gamma ray peak and usually an asymmetric sonic log signature which ranges from low velocities at the base to a higher velocity towards the top corresponding to the small-scale, gradual upward increase in limestone interbeds.

In the Northern Province (Figs 40, 41), the pink porcellaneous limestone at the base called the 'Paradoxica Bed' overlies the Red Chalk. The 'Paradoxica Bed' gives rise to the immediately overlying gamma ray low and high velocity sonic log spike. The remainder of the sequence comprises alternating marls and more or less hard chalks which give rise to a serrated, albeit subdued gamma ray curve and a more prominently serrated sonic log trace. Individual hard chalks and softer marls can be identified readily (Figs, 41, 42). As in the Southern Province, an overall upwards decrease in gamma ray values and increase in sonic velocity characterizes the upper part of the Lower Chalk. The sequence is capped by the Black Band which generates a prominent gamma ray peak and corresponding low velocity spike analogous to its partial correlative in the south, the Plenus Marls.

The Middle Chalk

The Melbourn Rock, marking the base of the Middle Chalk in the Southern Province (Fig. 40), has a characteristic geophysical log signature of very high sonic velocity and very low gamma ray values contrasting markedly with the underlying Plenus Marls. Towards the top of the Middle Chalk, however, flint courses occur which give rise to low gamma ray values and correspondingly high

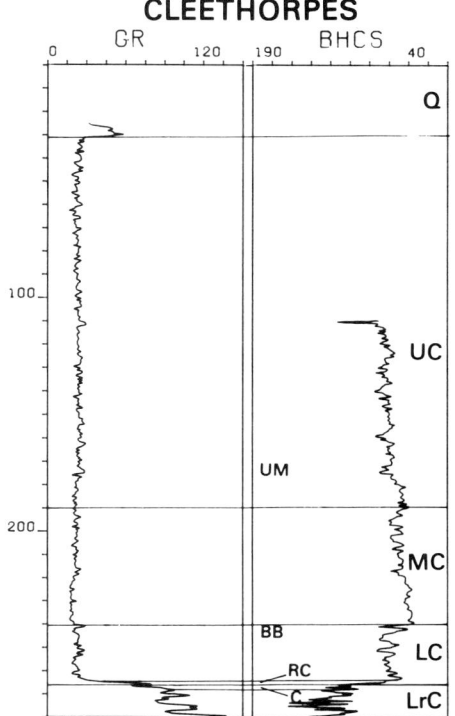

FIG. 41. Upper Cretaceous log signatures. LrC, Lower Cretaceous; C, Carstone; LC, Lower Chalk; RC, Red Chalk; BB, Black Band; MC, Middle Chalk; UC, Upper Chalk; UM, Ulceby Marl; Q, Quaternary.

sonic velocity and resistivity values. Since log suites in the Chalk tend to be limited it is impossible to distinguish these from the harder chalk levels. Like the Melbourn Chalk, these harder chalk levels are condensed 'hardground' deposits and have similar geophysical log characteristics. They are frequently of localized occurrence however, and that marking the top of the Middle Chalk, the 'Chalk Rock' (though belonging to the Upper Chalk), is less prominent in some places than an underlying 'Spurious Chalk Rock' which gives rise to a more prominent high velocity sonic spike. A prominent feature of the Middle Chalk is the occurrence of well-defined marl levels giving rise to minor gamma ray peaks and low values of sonic velocity and resistivity. These are extremely widespread but, as with the hardground levels, are condensed or missing in places and therefore their considerable value in regional correlation is locally reduced and sole dependence on the geophysical log signature is liable to be misleading.

The geophysical log signature of the Middle Chalk of the Northern Province (Figs 40, 41)

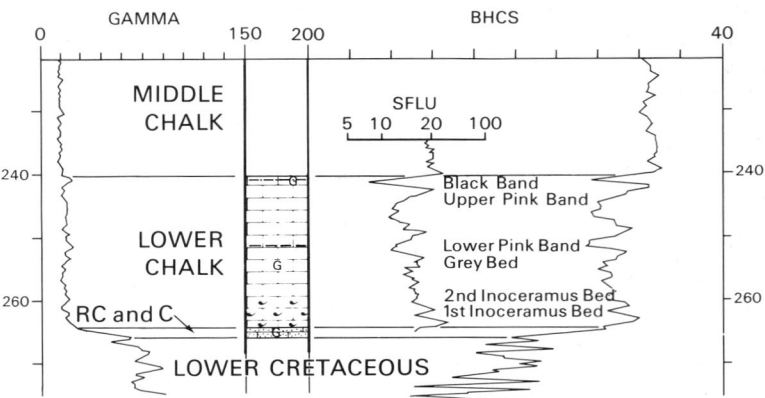

FIG. 42. Upper Cretaceous log signatures: Detail of Lower Chalk in the Cleethorpes Borehole. RC & C, Red Chalk and Carstone; SFLU, Spherically focused induction log.

resembles that of its southern correlative. Low gamma ray values and a corresponding sonic velocity peak mark the hardground complex at the base. Generally, lower sonic velocities characterize the overlying chalks where low velocity and resistance values and minor gamma ray peaks reflect conspicuous marl levels. In the upper part prominent sonic velocity spikes and a slight overall rise in sonic velocity reflect the fact that flint courses are interbedded with harder chalks from which they again may be difficult to distinguish on the basis of the log character of the usual restricted log suite. The Middle Chalk marls are very extensive laterally, though (as in the Southern Province) attenuation or disappearance associated with hardground development may result in local miscorrelation.

The Upper Chalk

The Upper Chalk of the Southern Province (Fig. 40) has a gamma ray signature characterized by low values. The soft, white chalks have a lower sonic velocity than those of the Lower and Middle Chalk although the nodular and tabular flints and hard chalks tend to give rise to a serrated log signature. Higher sonic velocities similar to those of the underlying Middle Chalk are usually obtained from the nodular beds in the basal part which are usually sufficiently prominent to merit the name Chalk Rock. Distinctive marl bands giving rise to minor gamma ray peaks and low sonic velocities are much less common than in underlying beds but do form distinctive marker beds of widespread occurrence. It is possible, too, to recognize distinctive flint seams and nodular hardgrounds though the latter, in particular, are localized and may not give a similar signature over large areas.

The overall signature of the Upper Chalk of the Northern Province (Figs 40, 41) resembles that of the Southern Province. Hardground chalks at the base give rise to low gamma ray values and high sonic velocities but generally the sonic velocity decreases upward as softer chalks dominate the sequence. High sonic velocities caused by harder chalks and flint courses also give the sonic log a serrated profile and conspicuous marl beds are present. These are as widespread as earlier marls but fewer in number. Again, individual hard chalks, flints and marl bands may be identified but these fine subdivisions have yet to be traced laterally.

Tertiary log signatures

Tertiary sediments in onshore Britain are confined to discrete sedimentary basins which owe their present disposition to the effects of mid-Tertiary tectonism. These dominantly terrigenous, and often poorly compacted, sediments lie with varying degrees of unconformity on lithologically contrasting strata in southern and western Britain. There is an overall E–W passage from dominantly marine sedimentation in south-eastern Britain, through interbedded, marine, brackish and non-marine sequences in central southern England to wholly non-marine, alluvial sequences in western Britain. Up to 250 m of late Palaeocene to late Eocene, dominantly marine, strata occur in the London Basin, which is the landward extension of the huge North Sea Basin. Some 600 m of late Palaeocene to mid-Oligocene marine, brackish and non-marine strata occur in the Hampshire Basin which extends south-eastwards into the cross-Channel, Dieppe

Basin. Farther west, in the fault-controlled Petrockstow and Bovey basins, between 600 m and 1200 m of non-marine sediments thought to be of late Eocene to Oligocene age occur. Similar sediments are found in isolated basins in the western British Isles as for example the 300 – ?1000 m sequence of Lough Neagh Clays situated in a downwarp in the Antrim Plateau or the 520 m of strata encountered in the Llandbedr (Mochras Farm) Borehole.

Stratigraphical correlation between separated basins is difficult since each contains widely differing rock-types and there are no marine fossils in the west. It is further complicated, however, by the presence of variously developed sedimentary cycles. Typically, in the east, basal glauconitic sands commonly rest on an erosion surface and pass up into clays which may themselves pass up into brackish and non-marine sands: to the west the higher parts of the cycle dominate and in the continental deposits of western Britain the cyclic sequence is related to local basin development and fluviatile sedimentation. In addition correlation is further complicated by the diachronous eastward spread of the less marine westerly facies.

The Mochras and Petrockstow basins

Full suites of geophysical logs are available for the cored boreholes which penetrated the full sequences of the non-marine basins at Mochras and Petrockstow (Masson-Smith 1971, Freshney *et al.* 1979). The boreholes penetrated similar geological successions comprising silty ligniferous clays with minor amounts of poorly compacted sand, silts and conglomerates. In both basins there is an overall increase in the proportion of coarser sediment towards the base. In both, too, fining-upward cycles are present on smaller and larger scales and have been interpreted as indicating channel-fill and overbank deposits of meandering river systems controlled by the periodic rejuvenation of nearby or subjacent fault-lines (O'Sullivan 1979, Freshney 1970). The extensional tectonic setting envisaged for these basins is therefore consistent with their Eocene to Oligocene age (Curry *et al.* 1978).

At Mochras and Petrockstow sonic velocity and formation density very obviously show gradual increase with depth (Masson-Smith 1971, Fig. 3) (Fig. 43). Minor fluctuations of both these log traces indicate the presence of clays (relatively lower sonic velocity and formation density) or silts, sands and conglomerates (relatively higher sonic velocity and formation density). Such fluctuations are picked out more prominently by the gamma ray trace where the coarser sediments yield low values and the clays, the highest values. Particularly

sensitive in this regard, however, is the neutron log in which higher values indicate the coarser grained sediments and lower values the clays. Although the Petrockstow Borehole was sited in the axial part of the basin where there is a predominance of river channel deposits and consequently lignitic and seat-earth levels are rare, lignites do occur particularly in the upper part of the sequence. They show, as they do at Mochras where they are more prominent, relatively high gamma ray values no doubt reflecting the fact that they are usually intimately associated with illitic clay (O'Sullivan 1979, p. 4). An exception to the otherwise remarkably consistent pattern of log response outlined above is the log response of the basal conglomerate at Mochras. Here gamma ray values are high in the lowest part and this may indicate a concentration of radioactive material in the unique predominance of pebbles of igneous rock in these basal beds.

The occurrence of small-scale, upward fining sedimentary cycles in these beds has been well illustrated by Freshney (1970, and Freshney *et al.* 1979) whose examples are indicated in Fig. 43. The repetitive increasing gamma ray values, which illustrate these cycles, also suggest the presence of larger scale cycles of 50–100 m thickness. The beds between 564 m and 660.5 m (Fig. 43) provide a clear example of a phenomenon which appears also to be present at Mochras. Both are related presumably to local fault-line movement but it is not possible at present to see any comparable patterns between both basins.

The Hampshire Basin

It is unfortunate that, generally, only restricted logging suites have been run in the Hampshire Basin because the sediments exhibit considerable lateral variation (Figs 43, 44). On the other hand the entire sequence has been cored in BGS boreholes and these have provided invaluable standards for stratigraphic calibration. Neither the sonic log nor the formation density log shows the same gradual upwards decrease in sonic velocity and density as may be seen in the more westerly, non-marine basins discussed above, hinting at a more complex post-Oligocene history of the Hampshire Basin. The sporadic nature of the consolidation and cementation of the coarser sediments results in considerable variability in sonic log signature. The tendency for these poorly consolidated levels to 'wash-out' during drilling, particularly coring, results in sections of hole sufficiently over-gauge to affect the density log to such an extent as to render its readings unreliable. The gamma ray log has been most commonly obtained but comparison of Figs 43 and 44, for example,

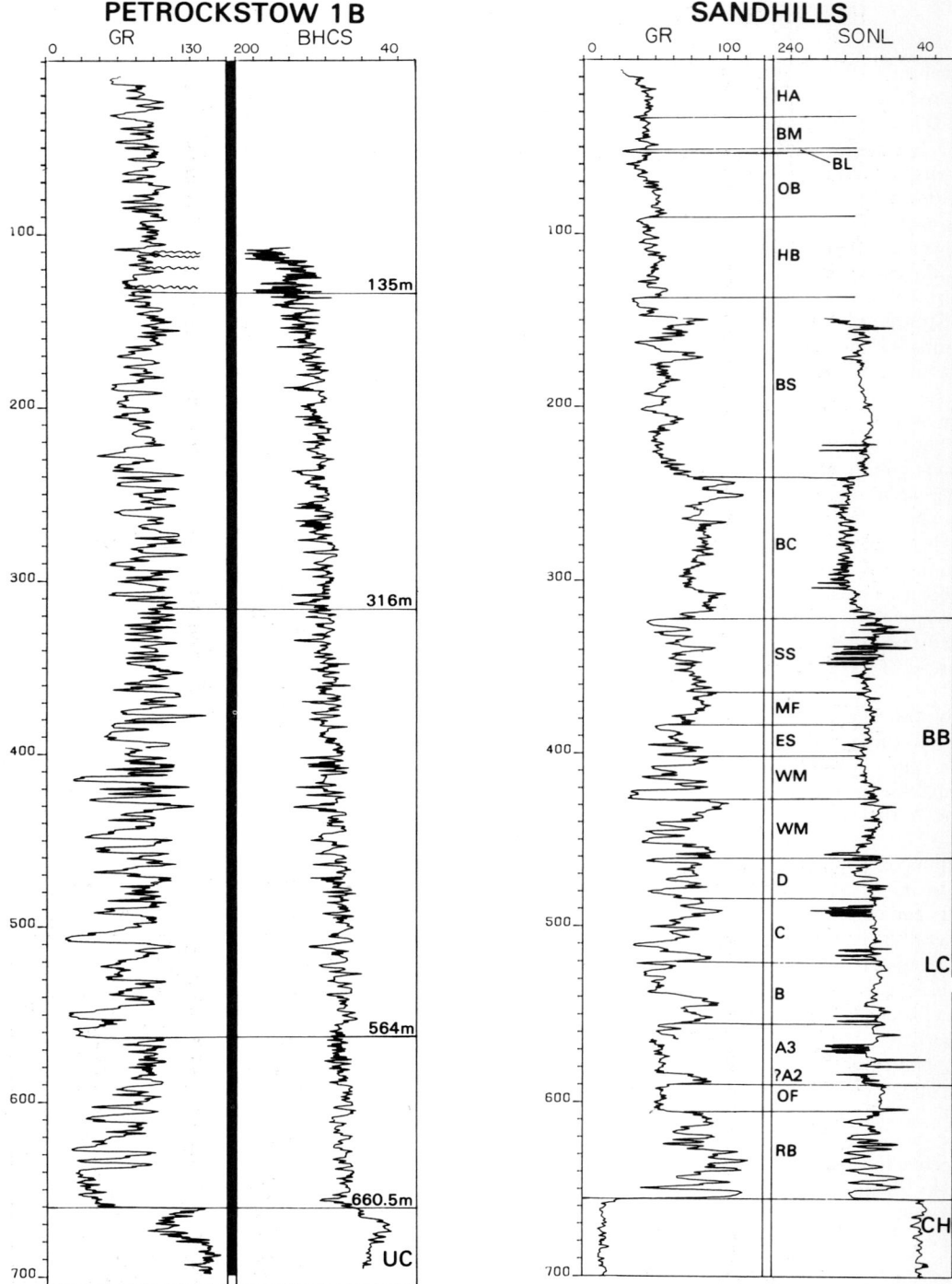

FIG. 43. Tertiary log signatures: south-west Britain and Hampshire Basin. Petrockstow 1B Borehole: UC, Upper Carboniferous; major subdivisions of Freshney et al. (1979) are shown by depth values; wavy lines on gamma ray trace indicate bases of the sedimentary cycles of Freshney et al. (1979, fig. 9). Sandhills Borehole: CH, Chalk; RB, Reading Beds; OF, Oldhaven Formation; LC, London Clay; A2, A3, B, C and D are sedimentary cycles of King (1981); BB, Bracklesham Beds; WM, Wittering Member; ES, Earnley Sand; MF, Marsh Farm Member; SS, Selsey Sand; BC, Barton Clay; BS, Barton Sand; HB, Headon Beds; OB, Osborne Beds; BL, Bembridge Limestone; BM, Bembridge Marl; HA, Hamstead Beds.

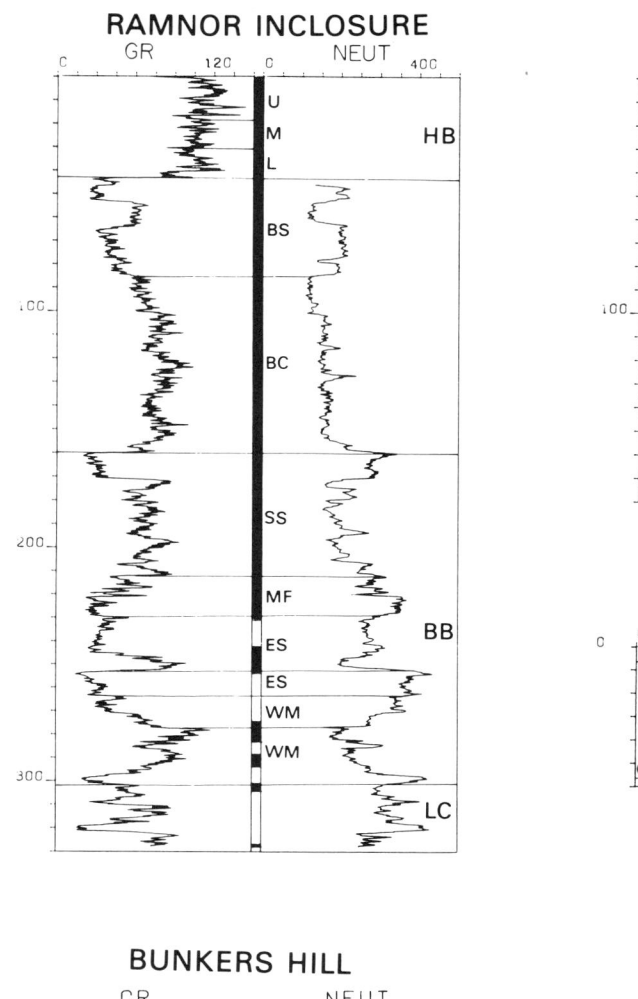

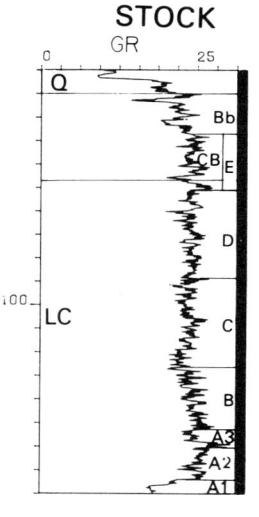

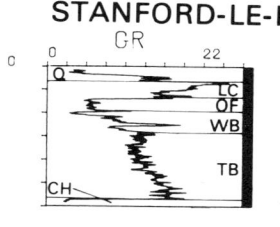

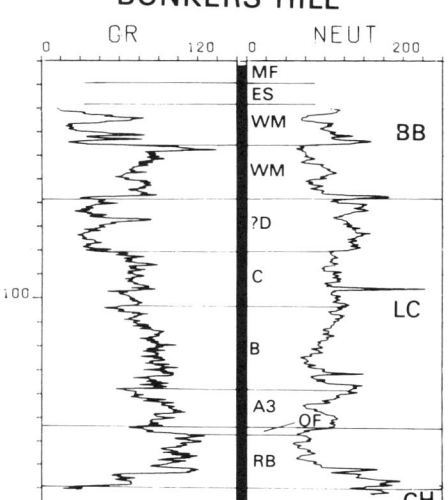

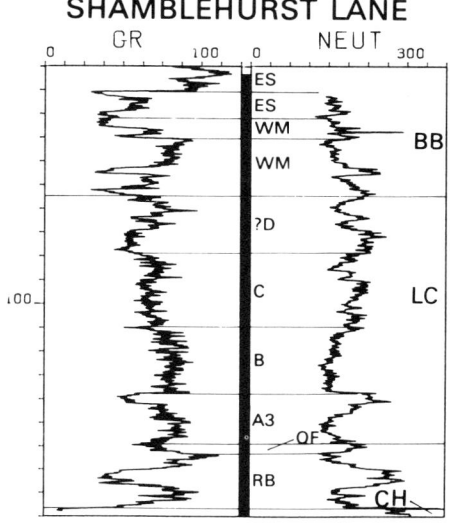

Fig. 44. Tertiary log signatures: London and Hampshire basins. CH, Chalk; RB, Reading Beds; TB, Thanet Beds; WB, Woolwich Beds; OF, Oldhaven Formation; LC, London Clay; A1, A2, A3, B, C, D and E are sedimentary cycles of King (1981); BB, Bracklesham Beds; WM, Wittering Member; ES, Earnley Sand; MF, Marsh Farm Member; SS, Selsey Sand; CB, Claygate Beds; Bb, Bagshot Beds; BC, Barton Clay; BS, Barton Sand; HB, Headon Beds including lower (L), middle (M) and upper (U) divisions; Q, Quaternary. All logs in counts/s.

shows the wide range of values encountered across the basin as a whole. Perhaps the next most useful log is the neutron trace which appears to respond well to the various sand bodies.

The soft, freshwater clays of the Reading Beds yield high gamma ray values and low sonic velocity and neutron values which contrast with those obtained from the harder Chalk beneath and the sandy Oldhaven Formation above. Discrete sandy horizons, however, give rise to lower gamma ray values, higher sonic velocity and neutron values. These alternations of sands and clays give the log motif a characteristic serrated appearance reminiscent of the log signatures of other brackish and freshwater sediments such as the Bathonian, Lower Estuarine Beds, Blisworth Clay and Forest Marble. Like those formations, too, individual features of the Reading Beds log traces cannot be widely stratigraphically correlated.

The geophysical log signature of the London Clay is best seen in the eastern and north-eastern parts of the Hampshire Basin (Fig. 44). It is characterized by an overall upward decrease in gamma ray values and decrease in neutron values reflecting an overall upward increase in sand content. These upward changes, however, are not gradual but cyclical and correspond to the succession of sedimentary cycles formalized by King (1981) who described units A to E in upward succession. Thus glauconitic sands to the base (the Oldhaven Formation of King 1981) give rise to low gamma ray but high neutron values. The sandy, basal part of the overlying London Clay proper becomes muddier upwards and this is reflected in the upward increasing gamma ray values (and decreasing neutron values). These trends are then reversed as the glauconitic sand content increases upwards again to reach a maximum at the top of Unit A (Fig. 44). A similar log motif characterizes the overlying Units B and C though here the clay-rich portion of the sequence is more prominent. Each of these major units comprises similar smaller units but generally, in the Hampshire Basin, the low gamma ray values (and high neutron values) to the top of the London Clay so predominate, reflecting the increased sandiness at the top of the formation, that the boundaries between Units C, D and E are not obvious. Elsewhere (e.g. Fig. 43) sands dominate the London Clay sequence such that recognition of cyclical sedimentary units is difficult and the overall upward change in motif described above is not readily apparent.

The alternating clays, silty clays and sands of the Bracklesham Beds are well shown by the alternating high and low gamma ray values of the geophysi-cal log traces (Figs 43, 44). Detailed elucidation is more difficult, however, for each of the clays and sands may show the effects of deposition in 'marine' conditions (e.g. the Earnley Sand and the Selsey Sand) or in 'reduced marine' conditions (e.g. the Wittering Member and the Marsh Farm Member) which have formed the basis of stratigraphical classification. Such conditions have been recognized in cored boreholes (Edwards & Freshney 1985) but comparison of these and other sections (Fig. 44) shows that there is no simple way of distinguishing individual units using the limited geophysical log data currently available. Indeed the presence of thicker 'non-marine' strata and thinner 'marine' strata to the west and south-west of the Hampshire Basin suggest that their mutual boundaries may be diachronous.

The Wittering Member (Figs 43, 44) is distinguished, in its lower part, by an intermittent upwards increase in gamma ray values (and corresponding decrease in neutron values) reflecting the upward passage from sands to clays. Where sandy horizons are developed to the top of the underlying London Clay, however, the boundary between Bracklesham Beds and London Clay may be difficult to draw. The gamma ray peak at the top of the lower part of the Wittering Member is widespread and marks the position of persistent ligniferous and paludal beds (Edwards & Freshney 1985). Above this horizon the gamma ray values decrease gradually or rapidly to persistent low values as the clays become silty and sandy to the top of the member. Such low values persist into the overlying Earnley Sand. Though the latter is glauconitic, in contrast to the sands of the underlying Wittering Member, it cannot be readily distinguished on the basis of the present geophysical log suites alone. As with the underlying member the Earnley Sand signature is tripartite overall. Lower and upper parts yield low gamma ray values (and higher neutron values) with a more or less well developed high gamma value peak in the middle to lower part of the member, reflecting a similar, albeit 'marine' sand, clay, sand sequence. Low gamma ray values and higher neutron values persist into the overlying 'reduced marine' sandy beds of the Marsh Farm Member. Again the basal parts of the latter cannot be readily differentiated on log character alone from the glauconitic sands of the Earnley Sand on which they lie. The Marsh Farm Member (Figs 43, 44) does appear to be characterized, however, by an upward increase in gamma ray values (and corresponding decrease in neutron values) marking its upward increasing clay content. The overlying Selsey Sand reverses these trends as it comprises an upward passage from clays to sands which is often more or less well cemented.

The overlying Barton Beds are easily recognized by the sharp change to lower gamma ray values, sonic velocity and neutron values which characterize this thick lower unit, the Barton Clay. A shift to lower gamma ray values, higher sonic velocity and neutron values characterizes the overlying Barton Sand, though minor clay horizons within the latter give rise to higher gamma ray (and lower neutron) values (Figs 43, 44). The distinctive bipartite signature of the Barton Beds is easy to recognize, partly because of the thickness of the units and also because they are usually sharply demarcated from contiguous strata. However, it would appear that the lowest part of the Barton Clay as used here is the correlative of the clays regarded as part of the Bracklesham Beds in the type area (see Edwards & Freshney 1985). The remainder of the Tertiary strata in the Hampshire Basin is not well known by log character (Figs 43, 44) though this is simply a result of the few logs that have been taken in boreholes.

The London Basin

Usually only gamma ray logs are available with which to characterize the Tertiary strata of the London Basin. At Stanford-le-Hope (Fig. 44), the Thanet Beds rest with marked unconformity on the Chalk as is reflected in the dramatic rise in gamma ray values at the base of the Tertiary sequence. Gamma ray values gradually decrease upwards as the Thanet Beds succession passes up from more or less glauconitic silt to very fine-grained sand. A minor increase in gamma ray values occurs to the top of the formation and it may be that this is associated with the diagenetic alteration of the topmost beds (Morton 1982, p. 270) prior to deposition of the Woolwich Beds. Notwithstanding this break in the sequence, the Woolwich Beds continue the trend to upward decreasing gamma ray values begun in the Thanet Beds. A similar profile may be seen at Canvey Island (Smart *et al.* 1964). To the west, however, the argillaceous upper part of the Woolwich Beds gives rise to prominent high gamma ray values overlying lower values which respond to the sandy lower part (Gray 1965). As would be expected, the glauconitic sands and silts of the Oldhaven Formation yield, on average, low gamma ray values whereas the overlying London Clay is marked by persistent, high gamma ray values (Fig. 44). Similar small-scale cycles to those recognized in the Hampshire Basin occur though the low gamma ray values associated with the presence of sands are not as prominent. Units A, B and C (King 1981, Fig. 13) are all recognizable

at Stock (Fig. 44) but Units D and E maintain high gamma ray values overall as the clays persist upwards. Silty levels are present and are marked by low gamma ray values. The latter increase somewhat in Unit E, most of which comprises the silty Claygate Beds, but there is no overall upward decrease in gamma ray values at this eastern end of the London Basin to compare with the overall asymmetric gamma ray curve encountered in sections farther west.

It is possible to interpret the uncored Canvey Island Borehole (Smart *et al.* 1964) in the light of these data such that the higher gamma ray values to the base of the London Clay (which overlie the prominent low values identified with the Blackheath Beds) may correspond to Units A1 and A2 of King (1981). The remainder of the London Clay signature resembles that of Stock (Fig. 44) where more prominent low gamma ray values occur in the overlying Bagshot Beds and reflect its interbedded sequence of sands, silts and grits.

ACKNOWLEDGEMENTS: This special report is published with the approval of the Director, British Geological Survey (NERC). The authors wish to acknowledge the assistance of their colleagues in the Deep Geology Research Group on whose interpretational experience they have freely drawn. In particular Dr C. J. Evans, Mr I. M. Burns and Mr A. G. Hulbert have helped with the preparation of numerous text figures. The account of compaction and depth of burial is based upon an unpublished report written by Dr G. A. Kirby and Mrs J. M. Allsop. We thank Dr R. N. Mortimore for allowing access to his unpublished work on the Chalk of southern England. The permission of Gas Council (Exploration) Limited to publish Fig. 43 is gratefully acknowledged.

REFERENCES

AITKENHEAD, N. 1977. The Institute of Geological Sciences Borehole at Duffield, Derbyshire. *Bull. geol. Surv. G.B.* **59**, 1–38.

ALLAUD, L. A. & MARTIN, M. 1977. *Schlumberger, the History of a Technique.* Wiley, London, 333 pp.

ALLSOP, J. M. 1974. Geophysical survey at the Spilmersford Borehole, East Lothian, Scotland. *Bull. geol. Surv. G.B.*, **45**, 63–72.

AMBROSE, K. & IVIMEY–COOK, H. I. C. 1982. The Barby (IGS) Borehole, near Daventry, Northamptonshire. *Rep. Inst. geol. Sci. London*, **82/1**, 36–40.

ARKELL, W. J. 1956. *Jurassic Geology of the World.* Oliver & Boyd, Edinburgh and London, xv + 806 pp.

ASHTON, M. 1980. The stratigraphy of the Lincolnshire Limestone Formation (Bajocian) in Lincolnshire and Rutland (Leicestershire). *Proc. geol. Assoc. London*, **91**, 203–223.

BALCHIN, D. A. & RIDD, M. F. 1970. Correlation of the younger Triassic rocks across eastern England. *Q. J. geol. Soc. London*, **126**, 91–101.

BARKER, R. D., LLOYD, J. W. & PEACH, D. W. 1984. The use of resistivity and gamma logging in litho-stragraphical studies of the Chalk in Lincolnshire and South Humberside. *Q. J. eng. geol. London,* **17,** 71–80.

BRADSHAW, M. J. & PENNEY, S. R. 1982. A cored Jurassic sequence from north Lincolnshire, England: stratigraphy, facies analysis and regional context. *Geol. Mag.* **119,** 113–134.

BRENNAND, T. P. 1975. The Triassic of the North Sea. *In* WOODLAND, A. W. (ed.). *Petroleum and the Continental Shelf of North-West Europe, 1, Geology.* Applied Science Publishers Ltd, London, 295–310.

BRUNSTROM, R. G. W. 1963. Recently discovered oilfields in Britain. *Proc. 6th Wld Petrol. Cong., Frankfurt,* Section 1, 11–20.

BURGESS, I. C. & HOLLIDAY, D. W. 1974. The Permo-Triassic rocks of the Hilton Borehole, Westmorland. *Bull. geol. Surv. G.B.* **46,** 1–34.

CASEY, R. 1967. The position of the Middle Volgian in the English Jurassic. *Proc. geol. Soc. London,* **1640,** 128–133.

CAVE, R. 1977. Geology of the Malmesbury District. *Mem. geol Surv. G.B.* vii + 343 pp.

COLTER, V. S. 1978. Exploration for gas in the Irish Sea. *Geol. Mijnbouw.* **57,** 503–516.

—— & REED, G. E. 1980. Zechstein 2 Fordon Evaporites of the Atwick 1 Borehole, surrounding area of N. E. England and the adjacent southern North Sea. *In* FUCHTBAUER, H. & PERYT, T. (eds). The Zechstein Basin. *Contrib. Sedimentol.* **9,** 115–129.

COPE, J. C. W., GETTY, T. A., HOWARTH, M. K., MORTON, N. & TORRENS, H. S. 1980. A correlation of Jurassic rocks in the British Isles. Part One: Introduction and Lower Jurassic. *Spec. Rep. geol. Soc. London,* **14,** 73 pp.

COX, B. M. & GALLOIS, R. W. 1981. The stratigraphy of the Kimmeridge Clay of the Dorset type area and its correlation with some other Kimmeridgian sequences. *Rep. Inst. geol. Sci. London,* **80/4,** 1–44.

CURRY, D., ADAMS, C. G., BOULTER, M. G., DILLEY, F. C., EAMES, F. E., FUNNELL, B. M. & WELLS, M. K. 1978. A correlation of Tertiary rocks in the British Isles. *Spec. Rep. geol. Soc. London,* **12,** 72 pp.

DAVIES, A. 1974. The Lower Carboniferous (Dinantian) sequence at Spilmersford, East Lothian, Scotland. *Bull. geol. Surv. G.B.* **45,** 1–24.

DAY, E. C. H. 1863. On the Middle and Upper Lias of the Dorsetshire coast. *Q.J. geol. Soc. London,* **19,** 278–97.

DOWNING, R. A. & HOWITT, F. 1969. Saline groundwaters in the Carboniferous rocks of the English East Midlands in relation to the geology. *Q.J. eng. Geol. London,* **1,** 241–269.

EDWARDS, R. A. & FRESHNEY, E. C. 1985. The geology of the country around Southampton. *Mem. geol. Surv. G.B.* (in press).

EDWARDS, W. N. 1967. Geology of the country around Ollerton. *Mem. geol. Surv. G.B.* x + 297 pp.

ELLIOTT, R. E. 1961. The stratigraphy of the Keuper Series in southern Nottinghamshire. *Proc. Yorkshire geol. Soc.* **33,** 197–234.

FRESHNEY, E. C. 1970. Cyclical sedimentation in the Petrockstow Basin. *Proc. Ussher Soc.* **2,** 179–89.

—— BEER, K. E. & WRIGHT, J. E. 1979. Geology of the country around Chulmleigh. *Mem. geol. Surv. G.B.* viii + 69 pp.

FROST, D. V. & HOLLIDAY, D. W. 1980. Geology of the country around Bellingham. *Mem. geol. Surv. G.B.* 112 pp.

GALLOIS, R. W. 1973. Some detailed correlations in the Upper Kimmeridge Clay in Norfolk and Lincolnshire. *Bull. geol. Surv. G.B.* **44,** 63–75.

—— 1979. Geological investigation for the Wash Water Storage Scheme. *Rep. Inst. geol. Sci. London,* **78/19,** 1–74.

—— & COX, B. M. 1974. Stratigraphy of the Upper Kimmeridge Clay of the Wash Area. *Bull. geol. Surv. G.B.* **47,** 1–28.

—— & —— 1976. The stratigraphy of the Lower Kimmeridge Clay of eastern England. *Proc. Yorkshire geol. Soc.* **41,** 13–26.

—— & —— 1977. The stratigraphy of the Middle and Upper Oxfordian sediments of Fenland. *Proc. Geol. Assoc. London,* **88,** 207–228.

GAUNT, G. D., IVIMEY-COOK, H. I. C., PENN, I. E. & COX, B. M. 1980. Mesozoic rocks proved by I.G.S. boreholes in the Humber and Acklam areas. *Rep. Inst. geol. Sci. London,* **78/13,** 1–34.

GEIGER, M. E. & HOPPING, C. A. 1968. Triassic stratigraphy of the southern North Sea Basin. *Phil. Trans. R. Soc. London,* **B254,** 1–36.

GEORGE, T. N., JOHNSON, G. A. L., MITCHELL, M., PRENTICE, J. E. RAMSBOTTOM, W. H. C., SEVASTOPULO, G. D. & WILSON, R. B. 1976. A correlation of Dinantian rocks in the British Isles. *Spec. Rep. geol. Soc. London,* **7,** 87 pp.

GRAY, D. A. 1958. Electrical resistivity marker bands in the Lower and Middle Chalk of the London Basin. *Bull. geol. Surv. G.B.* **15,** 85–95.

—— 1965. The stratigraphical significance of electrical resistivity marker bands in the Cretaceous strata of the Leatherhead (Fetcham Mill) Borehole, Surrey. *Bull. geol. Surv. G.B.* **23,** 65–115.

GREEN, G. W. & MELVILLE, R. V. 1956. The stratigraphy of the Stowell Park Borehole (1949–51). *Bull. geol. Surv. G.B.* **11,** 1–66.

HEMINGWAY, J. E. & RIDDLER, G. P. 1982. Basin inversion in North Yorkshire. *Trans. Instn. Ming. Metall.* **B91,** 6 175–186.

HORTON, A. & POOLE, E. G. 1977. The litho-stratigraphy of three geophysical marker horizons in the Lower Lias of Oxfordshire. *Bull. geol. Surv. G.B.* **62,** 13–24.

HOWARTH, M. K. 1957. The Middle Lias of the Dorset Coast. *Q.J. geol. Soc. London,* **113,** 185–204.

HOWITT, F. & BRUNSTROM, R. G. W. 1966. The continuation of the East Midlands Coal Measures into Lincolnshire. *Proc. Yorkshire. geol. Soc.* **35,** 549–64.

IVIMEY-COOK, H. C. & DONOVAN, D. T. 1983. The fauna of the Lower Jurassic. *In* WHITTAKER, A. & GREEN, G. W. Geology of the country around Weston-super-Mare. *Mem. geol. Surv. G.B.* 126–130.

KENT, P. E. 1980. Subsidence and uplift in east Yorkshire and Lincolnshire: a double inversion. *Proc. Yorkshire geol. Soc.* **42,** 505–524.

KING, C. 1981. The stratigraphy of the London Clay and associated deposits. *Tertiary Research Special Paper,* **6,** 158 pp. W. Backhuys, Rotterdam.

KIRBY, G. A., PENN, I. E. & SMITH, I. F. 1985. Cleethorpes No. 1: Geological Well Completion Report. *Invest. Geotherm. Potent. U.K., Rep. Br. geol. Surv.*

KNOWLES, B. 1964. The radioactive content of the Coal Measures sediments in the Yorkshire-Derbyshire Coalfield. *Proc. Yorkshire. geol. Soc.* **34**, 413–450.

LEE, M. K. 1984. Analysis of geophysical logs from the Shap, Skiddaw, Cairngorm, Ballater and Bennachie heat flow boreholes. *Invest. Geotherm. Potent. U.K., Rep. Br. geol. Surv.*

LOTT, G. K., SOBEY, R. A., WARRINGTON, G. & WHITTAKER, A. 1982. The Mercia Mudstone Group (Triassic) in the western Wessex Basin. *Proc. Ussher Soc.* **5**, 340–346.

McCANN, D., BARTON, K. J. & HEARN, K. 1981. Geophysical borehole logging with special reference to Altnabreac, Caithness. *Rep. Inst. geol. Sci. London. ENPU*, 81–11.

MAGARA, K. 1976. Thickness of removed sedimentary rocks, palaeopore pressure and palaeotemperature, southwestern part of West Canada Basin. *Bull. Am. Assoc. Petrol. Geol.* **60**, 554–565.

MARIE, J. P. P. 1975. Rotliegendes stratigraphy and diagenesis. *In* WOODLAND, A. W. (ed.). *Petroleum and the Continental Shelf of north west Europe. 1. Geology*. Applied Science Publishers, Barking, pp. 205–210, ix + 501pp.

MASSON-SMITH, D. 1971. Geophysical Investigations in the Llanbedr (Mochras Farm) Borehole. *Rep. Inst. geol. Sci. London*, **71/18**, 109–111.

MORTIMORE, R. N. 1983. The stratigraphy and sedimentation of the Turonian-Campanian in the Southern Province of England. *Zitteliana*, **10**, 27–41.

MORTON, A. C. 1982. The provenance and diagenesis of Palaeogene sandstones of south-east England as indicated by their heavy mineral analysis. *Proc. geol. Assoc. London*, **93**, 263–274.

MURRAY, K. H. 1982a. Correlation of electrical resistivity marker bands in the Chalk of the London Basin. *Hydrogeol. Rep., Inst. geol. Sci. London.*, **WD/82/1**, 1–6.

—— 1982b. Correlation of electrical resistivity marker bands in the Cenomanian and Turonian Chalk from the London Basin to the Market Weighton area, Yorkshire. *Hydrogeol. Rep., Inst. geol. Sci. London.*, **WD/82/5**, 1–12.

O'SULLIVAN, K. N. 1979. The sedimentology, geochemistry and conditions of deposition of the Tertiary rocks of the Llanbedr (Mochras Farm) Borehole. *Rep. Inst. geol. Sci. London*, **78/24**, 1–13.

PALMER, C. P. 1972. The Lower Lias (Lower Jurassic) between Watchet and Lilstock in North Somerset (United Kingdom). *Newsl. Stratig.*, **2**, 1–30.

PARRY, G. C., WHITLEY, P. K. J. & SIMPSON, R. D. H. 1981. Integration of palynological and sedimentological methods in facies analysis of the Brent Formation. *In* ILLING, L. V. & HOBSON, G. D. (eds). *Petroleum Geology of the Continental Shelf of north-west Europe*. Heyden & Son Ltd and Institute of Petroleum, London, pp. 205–215, xvii + 521pp.

PENN, I. E. 1982. Middle Jurassic stratigraphy and correlation of the Winterborne Kingston borehole, Dorset. *In* RHYS, G. H., LOTT, G. K. & CALVER, M. A. (eds). The Winterborne Kingston Borehole, Dorset, England. *Rep. Inst. geol. Sci. London*, **81/3**, 53–76.

——, HOLLIDAY, D. W., KIRBY, G. A., KUBALA, M., SOBEY, R. A., MITCHELL, W. I., HARRISON, R. K. & BECKINSALE, R. D. 1983. The Larne No. 2 Borehole: discovery of a new Permian volcanic centre. *Scott. J. Geol.*, **19**, 333–346.

——, MERRIMAN, R. J. & WYATT, R. J. 1979. The Bathonian strata of the Bath-Frome area. *Rep. Inst. geol. Sci. London*, **78/22**, 1–88.

PONSFORD, D. R. A. 1955. Radioactivity studies of some British sedimentary rocks. *Bull. geol. Surv. G.B.* **10**, 24–44.

POOLE, E. G. 1969. The stratigraphy of the Geological Survey Apley Barn Borehole Witney, Oxfordshire. *Bull. geol. Surv. G.B.* **29**, 1–104.

—— 1977. Stratigraphy of the Steeple Aston Borehole. *Bull. geol. Surv. G.B.* **57**, 1–85.

—— 1978. Stratigraphy of the Withycombe Farm Borehole, near Banbury, Oxfordshire. *Bull. geol. Surv. G.B.* **68**, 1–63.

POWELL, J. H. 1984. Lithostratigraphical nomenclature of the Lias Group in the Yorkshire Basin. *Proc. Yorkshire geol. Soc.*, **45**, 51–57.

RAMSBOTTOM, W. H. C., RHYS, G. H. & SMITH, E. G. 1962. Boreholes in the Carboniferous rocks of the Ashover district, Derbyshire. *Bull. geol. Surv. G.B.* **19**, 75–168.

RAWSON, P. F., CURRY, D., DILLEY, F. C., HANCOCK, J. M., KENNEDY, W. J., NEALE, J. W., WOOD, C. J. & WORSSAM, B. C. 1978. A correlation of Cretaceous rocks in the British Isles. *Spec. Rep. geol. Soc. London*, **9**, 70 pp.

RIDD, M. F., WALKER, D. B. & JONES, J. M. 1970. A deep borehole at Harton on the margin of the Northumbrian Trough. *Proc. Yorkshire. geol. Soc.* **38**, 75–103.

SCHLUMBERGER 1979. *Log Interpretation Charts*. Schlumberger Ltd, USA, iv + 97 pp.

SELLEY, R. C. 1976. Subsurface environmental analysis of North Sea sediments. *Bull. Am. Assoc. Petrol. Geol.* **60**, 184–195.

SMART, J. G. O., SABINE, P. A. & BULLERWELL, W. 1964. The Geological Survey exploratory borehole at Canvey Island, Essex. *Bull. geol. Surv. G.B.* **21**, 1–36.

SMITH, D. B. & CROSBY, A. 1979. The regional and stratigraphical context of Zechstein 3 and 4 potash deposits in the British sector of the Southern North Sea and adjoining land areas. *Econ. Geol.* **74**, 397–408.

SMITH, E. G., RHYS, G. H. & GOOSENS, R. F. 1973. Geology of the country around East Retford, Worksop and Gainsborough. *Mem. geol. Surv. G.B.* XI + 348 pp.

TAYLOR, J. C. M. 1980. Origin of the Werraanhydrit in the U.K. Southern North Sea — a reappraisal. *In* FUCHTBAUER, H. & PERYT, T. (eds). The Zechstein Basin. *Contri. Sedimentol.* **9**, 91–113.

—— 1981. Zechstein facies and petroleum prospects in the central and northern North Sea. *In* ILLING, L. V. & HOBSON, G. D. (eds). *Petroleum Geology of the Continental Shelf of north-west Europe*. Heyden & Son Ltd and Institute of Petroleum, London, pp. 176–185, xviii + 521 pp.

TAYLOR, J. C. M. & COLTER, V. S. 1975. Zechstein of the English sector of the southern North Sea Basin. *In* WOODLAND, A. W. (ed.). *Petroleum and the Continental Shelf of North-West Europe. 1. Geology.* Applied Science Publishers, Barking, pp. 249–263, ix + 501 pp.

THOMAS, L. P. & HOLLIDAY, D. W. 1982. Southampton No. 1 (Western Esplanade) Geothermal Well: Geological Well Completion Report. *Rep. Deep. Geol. Unit, Inst. geol. Sci. London,* **82/3,** 1–41.

TUNBRIDGE, I. P. 1983. Geophysical down-hole recognition of the Lower Devonian *'Psammosteus'* Limestone and Townsend Tuff Bed, South Wales. *Geol. J.* **18,** 325–329.

WARRINGTON, G., AUDLEY-CHARLES, M. E., ELLIOTT, R. E., EVANS, W. B., IVIMEY-COOKE, H. C., KENT, P., ROBINSON, P. L., SHOTTON, F. W. & TAYLOR, F. M. 1980. A correlation of Triassic rocks in the British Isles. *Spec. Rep. geol. Soc. London,* **13,** 78 pp.

WHITTAKER, A. 1980. Shrewton No. 1: Geological well completion report. *Rep. Deep Geol. Unit, Inst. geol. Sci. London,* **80/1,** 1–34.

———— 1980b. Kempsey No. 1: Geological well completion report. *Rep. Deep Geol. Unit. geol. Sci. London,* **80/1,** 1–14.

———— 1980c. Marchwood No. 1 Geological Well Completion Report. *Rep. Deep. Geol. Unit, Inst. geol. Sci. London,* **80/5,** 1–55.

———— & GREEN, G. W. 1983. Geology of the country around Weston-super-Mare. *Mem. geol. Surv. G.B.,* Sheet 279 with parts of 263 and 295, 147 pp.

WOOD, C. J. & SMITH, E. G. 1978. Lithostratigraphical classification of the Chalk in north Yorkshire, Humberside and Lincolnshire. *Proc. Yorkshire geol. Soc.* **42,** 263–287, pls 11–12.

WORSSAM, B. C. & IVIMEY-COOKE, H. I. C. 1971. The stratigraphy of the Geological Survey borehole at Warlingham, Surrey. *Bull. geol. Surv. G.B.* **36,** 1–111, pls 1–3.

A. WHITTAKER, D. W. HOLLIDAY AND I. E. PENN, Deep Geology Research Group, British Geological Survey, Keyworth, Nottingham NG12 5GG.